Advanced DSP Techniques for High-Capacity and Energy-Efficient Optical Fiber Communications

Advanced DSP Techniques for High-Capacity and Energy-Efficient Optical Fiber Communications

Special Issue Editors

Zhongqi Pan
Yang Yue

MDPI • Basel • Beijing • Wuhan • Barcelona • Belgrade

Special Issue Editors
Zhongqi Pan
University of Louisiana at Lafayette
USA

Yang Yue
Nankai University
China

Editorial Office
MDPI
St. Alban-Anlage 66
4052 Basel, Switzerland

This is a reprint of articles from the Special Issue published online in the open access journal *Applied Sciences* (ISSN 2076-3417) from 2018 to 2019 (available at: https://www.mdpi.com/journal/applsci/special_issues/DSP_Optical_Fiber_Communication).

For citation purposes, cite each article independently as indicated on the article page online and as indicated below:

LastName, A.A.; LastName, B.B.; LastName, C.C. Article Title. *Journal Name* **Year**, *Article Number*, Page Range.

ISBN 978-3-03921-792-2 (Pbk)
ISBN 978-3-03921-793-9 (PDF)

Contents

About the Special Issue Editors

Zhongqi Pan received B.S. and M.S. degrees from Tsinghua University, China, and a Ph.D. degree from the University of Southern California, Los Angeles, all in electrical engineering. He is currently a Professor at the Department of Electrical and Computer Engineering. He also holds BORSF Endowed Professorship in Electrical Engineering II, and BellSouth/BoRSF Endowed Professorship in Telecommunications. Dr. Pan's research is in the area of photonics, including photonic devices, fiber communications, wavelength-division multiplexing (WDM) technologies, optical performance monitoring, coherent optical communications, space-division multiplexing (SDM) technologies, and fiber-sensor technologies. He has authored/co-authored 160 publications, including five book chapters and >20 invited presentations/papers. He also has five U. S. patents and one Chinese patent. Prof. Pan is an OSA and IEEE senior member.

Yang Yue received B.S. and M.S. degrees in electrical engineering and optics from Nankai University, Tianjin, China, in 2004 and 2007, respectively. He received a Ph.D. degree in electrical engineering from the University of Southern California, Los Angeles, CA, USA, in 2012. He is a Professor with the Institute of Modern Optics, Nankai University, Tianjin, China. Dr. Yue's current research interests include intelligent photonics, optical communications and networking, optical interconnect, detection, imaging, and display technology. He has published over 150 peer-reviewed journal papers and conference proceedings, two edited books, one book chapter, >10 invited papers, >30 issued or pending patents, and >60 invited presentations.

Editorial

Special Issue on Advanced DSP Techniques for High-Capacity and Energy-Efficient Optical Fiber Communications

Zhongqi Pan [1] and Yang Yue [2,*]

[1] Department of Electrical and Computer Engineering, University of Louisiana at Lafayette, Lafayette, LA 70504, USA; zpan@louisiana.edu
[2] Institute of Modern Optics, Nankai University, Tianjin 300350, China
* Correspondence: yueyang@nankai.edu.cn; Tel.: +86-22-8535-8565

Received: 8 October 2019; Accepted: 18 October 2019; Published: 22 October 2019

1. Introduction

The rapid proliferation of the Internet has been driving communication networks closer and closer to their limits, while available bandwidth is disappearing due to ever-increasing network loads. In the past decade, optical fiber communication technology has increased the per fiber data rate from 10 Tb/s to over 10 Pb/s [1]. A major explosion came after the maturity of coherent detection and advanced digital signal processing (DSP), which enabled the achievement of high spectral and energy efficiency. It is difficult to overstate the impact that optical coherent technologies have had in both generating and supporting the revolution of optical communications over the last 10 years.

As one of the key enablers in the coherent evolution of fiber communication systems, DSP has made the innovation of high-order modulation formats possible in increasing spectral/power efficiency. DSP can also compensate almost all linear and nonlinear distortions, and improve noise tolerance in fiber systems. It provides a promising electrical solution for many problems in the optical domain. For example, troublesome channel impairments due to chromatic dispersion (CD) and polarization mode dispersion (PMD) can be almost fully compensated through DSP-based equalizers. Furthermore, DSP has the potential to ease the requirements for many optical components and can even take the place of some optical functions.

2. Special Issue Papers

In this Special Issue, we present nine carefully selected papers, including three review articles and six contributed papers. These papers cover advanced DSP techniques for long-distance, short-reach applications, including systems that use conventional single mode fibers (SMFs) and those based on space-division multiplexing (SDM) fibers, as well as links that use free-space wireless transmission.

The following summarizes these papers:

- Advanced DSP for Coherent Optical Fiber Communication [2]

  This paper provides an overview of recent progress on advanced DSP techniques for high-capacity, long-haul, coherent optical fiber transmission systems. The authors first introduce the principle and scheme of coherent detection to explain why DSP can compensate for transmission impairments. Then, the corresponding techniques for nonlinearity compensation, frequency-domain equalization (FDE), SDM, and machine learning (ML) are discussed. Relevant techniques are analyzed, and representational results and experimental verifications are demonstrated as well. This paper also provides a brief conclusion and future perspectives at the end.
- Recent Advances in DSP Techniques for Mode Division Multiplexing Optical Networks with MIMO Equalization: A Review [3]

This paper provides a technical review regarding the latest progress on multi-input multi-output (MIMO) DSP equalization techniques for high-capacity fiber-optic communication networks. The authors discuss state of the art of MIMO equalizers, predominantly focusing on the advantages of implementing the space–time block coding (STBC)-assisted MIMO technique. They also present a performance evaluation of different MIMO frequency-domain equalization (FDE) schemes for differential mode group delay (DMGD) and mode-dependent loss (MDL) issues in adaptive coherent receivers. Moreover, optimization of hardware complexity in MIMO-DSP is discussed, and a joint-compensation scheme is deliberated for CD and DMGD, along with a number of recent experimental demonstrations using MIMO-DSP.

- Recent Advances in Equalization Technologies for Short-Reach Optical Links Based on PAM4 Modulation: A Review [4]

  The authors review the latest progress on DSP equalization technologies for short-reach optical links based on four-level pulse amplitude modulation (PAM4) modulation. They introduce the configuration and challenges of the transmission system, and cover the principles and performance of different equalizers and some improved methods. In addition, machine learning algorithms are discussed to mitigate nonlinear distortion for next-generation short-reach PAM4 links. A summary of various equalization technologies is illustrated, and a perspective of the future trend is given as well.

- Experimental Investigation of 400 Gb/s Data Center Interconnect Using Unamplified High-Baud-Rate and High-Order QAM Single-Carrier Signal [5]

  The authors review the latest progress on data center interconnects (DCI). They also discuss different perspectives on the 400G pluggable module, including form factor, architecture, digital signal processing (DSP), and module power consumption, following 400G pluggable optics in DCI applications. The authors also experimentally investigate the capacity-reach matrix for high-baud-rate and high-order quadrature amplitude modulation (QAM) single-carrier signals in unamplified single-mode optical fiber (SMF) links.

- Compensation of Limited Bandwidth and Nonlinearity for Coherent Transponder [6]

  The authors present a novel solution for optimizing the coefficients of digital filters to mitigate impairments due to limited bandwidth and nonlinearity in coherent transponders. They show that limited bandwidth is improved by the finite impulse response filter, while nonlinearity is mitigated by the memoryless Volterra filter.

- Optical Transmitters without Driver Amplifiers-Optimal Operation Conditions [7]

  The authors discuss the influence of waveform design on the root-mean-square amplitude and the optical signal quality generated by a Mach–Zehnder modulator with a limited electrical swing (Vpp). Specifically, the influence of the pulse shape, clipping, and digital pre-distortion on the signal quality after the electrical-to-optical conversion are investigated. The findings are of interest for single-channel intensity modulation and direct detection (IM/DD) links, as well as optical coherent communication links.

- Performance Improvement for Mixed RF–FSO Communication System by Adopting Hybrid Subcarrier Intensity Modulation [8]

  This paper presents research on end-to-end mixed radio frequency–free space optical (RF–FSO) systems with the hybrid pulse position modulation–binary phase shift keying–subcarrier intensity modulation (PPM–BPSK–SIM) scheme in wireless optical communications. The RF link obeys Rayleigh distribution and the FSO link experiences gamma-gamma distribution. The average bit error rate (BER) for various PPM–BPSK–SIM schemes is derived with consideration of atmospheric turbulence influence and pointing error condition. The outage probability and the average channel capacity of the system are discussed as well.

- Post-FEC Performance of Pilot-Aided Carrier Phase Estimation over Cycle Slip [9]

  The authors present the post-forward error correction (FEC) bit error rate (BER) performance and the cycle-slip (CS) probability of the carrier phase estimation (CPE) scheme based on the Viterbi–Viterbi phase estimation (VVPE) algorithm and the VV cascaded by the pilot-aided-phase-unwrap (PAPU) algorithm in a 56 Gbit/s quadrature phase-shift keying (QPSK) coherent communication system.
- IBP Based Caching Strategy in D2D [10]

  Device-to-device (D2D) communication is a key technology in 5G wireless systems, increasing communication capacity and spectral efficiency. In this paper, the authors propose an Indian buffet process-based D2D caching strategy (IBPSC) to provide high quality D2D communications according to physical closeness between devices. Experimental results show that IBPSC achieves the best network performance.

3. Future Trend

Network traffic has been increasing exponentially over decades. This enormous growth rate will continue, in the foreseeable future, due to many newly-emerging and unanticipated digital applications and services in the 5G network. To fulfill the ever-growing bandwidth demand, not only do the spectral efficiencies of optical fiber communication systems need to be further improved, but also the power/wavelength needs to be reduced so that higher individual data rates per wavelength (up to multi-Tb/s) can be achieved with total aggregate capacities well beyond Pb/s. As one of the most prominent enabling technologies, DSP has played a critical role in accommodating channel impairment mitigation, enabling advanced modulation formats for spectral-efficient transmission, and realizing flexible bandwidth. We believe more innovations in DSP techniques are needed to further reduce the cost per bit, increase network efficiency, and approach the Shannon limit. We anticipate that more optical functionalities will be achieved by DSP in the electrical domain, making future communication networks more efficient and flexible.

Acknowledgments: First of all, the guest editors would like to thank all the authors for their excellent contributions to this special issue. Secondly, we would like to thank all the reviewers for their outstanding job in evaluating the manuscripts and providing valuable comments. Additionally, the guest editors would like to thank the MDPI team involved in the preparation, editing, and managing of this special issue. Finally, we would like to express our sincere gratitude to Lucia Li, the contact editor of this special issue, for her kind, efficient, professional guidance and support through the whole process. It would not be possible to have the above collection of high quality papers without these joint efforts.

Conflicts of Interest: The authors declare no conflict of interest.

References

1. Soma, D.; Wakayama, Y.; Beppu, S.; Sumita, S.; Tsuritani, T.; Hayashi, T.; Nagashima, T.; Suzuki, M.; Yoshida, M.; Kasai, K.; et al. 10.16-Peta-B/s Dense SDM/WDM Transmission Over 6-Mode 19-Core Fiber Across the C+L Band. *J. Lightwave Technol.* **2018**, *36*, 1362–1368. [CrossRef]
2. Zhao, J.; Liu, Y.; Xu, T. Advanced DSP for coherent optical fiber communication. *Appl. Sci.* **2019**, *9*, 4192. [CrossRef]
3. Weng, Y.; Wang, J.; Pan, Z. Recent Advances in DSP Techniques for Mode Division Multiplexing Optical Networks with MIMO Equalization: A Review. *Appl. Sci.* **2019**, *9*, 1178. [CrossRef]
4. Zhou, H.; Li, Y.; Liu, Y.; Yue, L.; Gao, C.; Li, W.; Qiu, J.; Guo, H.; Hong, X.; Zuo, Y.; et al. Recent Advances in Equalization Technologies for Short-Reach Optical Links Based on PAM4 Modulation: A Review. *Appl. Sci.* **2019**, *9*, 2342. [CrossRef]
5. Yue, Y.; Wang, Q.; Anderson, J. Experimental Investigation of 400 Gb/s Data Center Interconnect Using Unamplified High-Baud-Rate and High-Order QAM Single-Carrier Signal. *Appl. Sci.* **2019**, *9*, 2455. [CrossRef]
6. Wang, Q.; Yue, Y.; Anderson, J. Compensation of Limited Bandwidth and Nonlinearity for Coherent Transponder. *Appl. Sci.* **2019**, *9*, 1758. [CrossRef]

7. Josten, A.; Baeuerle, B.; Bonjour, R.; Heni, W.; Leuthold, J. Optical Transmitters without Driver Amplifiers—Optimal Operation Conditions. *Appl. Sci.* **2018**, *8*, 1652. [CrossRef]
8. Jiang, T.; Zhao, L.; Liu, H.; Deng, D.; Luo, A.; Wei, Z.; Yang, X. Performance Improvement for Mixed RF–FSO Communication System by Adopting Hybrid Subcarrier Intensity Modulation. *Appl. Sci.* **2019**, *9*, 3724. [CrossRef]
9. Li, Y.; Ning, Q.; Yue, L.; Zhou, H.; Gao, C.; Liu, Y.; Qiu, J.; Li, W.; Hong, X.; Wu, J. Post-FEC Performance of Pilot-Aided Carrier Phase Estimation over Cycle Slip. *Appl. Sci.* **2019**, *9*, 2749. [CrossRef]
10. Shan, C.; Wu, X.-P.; Liu, Y.; Cai, J.; Luo, J.-Z. IBP Based Caching Strategy in D2D. *Appl. Sci.* **2019**, *9*, 2416. [CrossRef]

Review

Advanced DSP for Coherent Optical Fiber Communication

Jian Zhao [1,*], Yaping Liu [1] and Tianhua Xu [1,2,*]

1 Key Laboratory of Opto-Electronic Information Technical Science of Ministry of Education, School of Precision Instruments and Opto-Electronics Engineering, Tianjin University, Tianjin 300072, China; liuyp@tju.edu.cn
2 School of Engineering, University of Warwick, Coventry CV4 7AL, UK
* Correspondence: enzhaojian@tju.edu.cn (J.Z.); tianhua.xu@ieee.org (T.X.)

Received: 27 August 2019; Accepted: 29 September 2019; Published: 8 October 2019

Abstract: In this paper, we provide an overview of recent progress on advanced digital signal processing (DSP) techniques for high-capacity long-haul coherent optical fiber transmission systems. Not only the linear impairments existing in optical transmission links need to be compensated, but also, the nonlinear impairments require proper algorithms for mitigation because they become major limiting factors for long-haul large-capacity optical transmission systems. Besides the time domain equalization (TDE), the frequency domain equalization (FDE) DSP also provides a similar performance, with a much-reduced computational complexity. Advanced DSP also plays an important role for the realization of space division multiplexing (SDM). SDM techniques have been developed recently to enhance the system capacity by at least one order of magnitude. Some impressive results have been reported and have outperformed the nonlinear Shannon limit of the single-mode fiber (SMF). SDM introduces the space dimension to the optical fiber communication. The few-mode fiber (FMF) and multi-core fiber (MCF) have been manufactured for novel multiplexing techniques such as mode-division multiplexing (MDM) and multi-core multiplexing (MCM). Each mode or core can be considered as an independent degree of freedom, but unfortunately, signals will suffer serious coupling during the propagation. Multi-input–multi-output (MIMO) DSP can equalize the signal coupling and makes SDM transmission feasible. The machine learning (ML) technique has attracted worldwide attention and has been explored for advanced DSP. In this paper, we firstly introduce the principle and scheme of coherent detection to explain why the DSP techniques can compensate for transmission impairments. Then corresponding technologies related to the DSP, such as nonlinearity compensation, FDE, SDM and ML will be discussed. Relevant techniques will be analyzed, and representational results and experimental verifications will be demonstrated. In the end, a brief conclusion and perspective will be provided.

Keywords: optical fiber communication; digital signal processing; coherent detection; equalization; nonlinearity compensation; space division multiplexing; machine learning; neural network

---

1. Introduction

With the development of Erbium doped fiber amplifier (EDFA), wavelength division multiplexing (WDM), dispersion management and optical fiber nonlinearity compensation technologies, optical fiber communication capacity has been rapidly improved over the past few decades. In addition, the research on high-order modulation formats has also increased the transmission capacity and spectral efficiency (SE) of optical fibers. Due to a smaller bandwidth occupied (for the same bit rate), systems with higher SE are usually more tolerant to chromatic dispersion (CD) and polarization mode dispersion (PMD). The fault tolerance to CD and PMD are particularly crucial in high bit-rate transmission systems. In order to achieve a high SE, early works used direct detection (incoherent detection) and amplitude-based modulation (1-dimension). Actually, higher SE can be obtained using coherent detection and 2-dimension modulation formats.

Coherent detection attracted intensive research in the 1980s due to its high receiver sensitivity. With the invention of EDFA and the development of WDM systems, research in coherent optical communication ceased in early 1990s, because of the difficulty and complexity of system implementation, especially the complicated realization of optical phase locked loop. Phase and polarization management turned out to be the major obstacles for the practical implementation of conventional coherent receivers. Fortunately, both phase and polarization management can be realized in the electrical domain using DSP. With the increasing demand on transmission capacity, coherent optical communication has attracted widespread attention again in recent years, and has become an unprecedented and promising approach for realizing high-capacity long-haul optical communication systems [1–5].

From the beginning of this century, coherent detection combined with large-bandwidth analog-to-digital converter (ADC), digital-to-analog converter (DAC) and DSP has increased the achievable capacity of optical fiber communication systems [6–10]. In coherent optical communication, information is encoded onto optical carrier waves in the complex domain, and the optical signal carrying all information is transmitted to the front end, after undergoing a series of linear and nonlinear impairments. In order to recover the transmitted signal and obtain the original information, it is necessary to measure the full complex electric field of the light wave, which means that the phase and the magnitude of optical carrier field both need to be detected. Using coherent detection, the complex field of the received signal can be completely acquired, and the linear transmission impairments, such as CD and PMD, can be fully compensated by using static and adaptive DSP.

One of the main benefits of coherent optical communication is the possibility of compensation of transmission impairments using DSP. Compensation of linear impairments such as CD and PMD are now implemented using DSP. Compensation of nonlinear impairments brings about an increase in the fundamental capacity limit for fiber transmission, which is an important research area for coherent optical communications. Advanced DSP has also been developed in the frequency domain to reduce the computational complexity and maintain all of the advantages of time domain equalization (TDE) based adaptive digital filters. Machine learning (ML) techniques have become one of the most promising disciplines and is beneficial to optical fiber communication applications such as nonlinearity mitigation, optical performance monitoring (OPM), carrier recovery, in-band optical signal-to-noise ratio (OSNR) estimation and modulation format classification, and especially, advanced DSP.

In this paper, the latest development and progress of DSP approaches for coherent optical communications are reviewed. The paper is organized as follows: Section 2 describes the principle and schematic of the coherent detection as well as linear impairments equalization. From Section 3 to Section 5, nonlinearity compensation, SDM applications and FDE approaches are presented in detail, respectively. Section 6 will explore the application of recently proposed and promising ML technologies in optical communications. A brief conclusion with our perspectives is provided in the last section.

2. Principle of Coherent Detection and Linear Impairments Equalization

The coherent receiver based on an intradyne system is shown in Figure 1a [11]. The input signal interferes with the local oscillator (LO) laser, which usually has the same frequency as the transmitter laser in the 90-degree optical hybrid device. Balanced detectors (BDs) are often used to reject the common mode noise. In order to detect both real and imaginary parts, the input signal is mixed with the real part of the LO in one arm and the imaginary part in the other arm through the 90-degree phase delay between the signal and the LO introduced by the 90-degree hybrid. Polarization diversity is introduced to ensure the detection of both polarizations of the signal. It is noted that the LO does not require the phase and polarization locking for the input signal. Then the electrical signal is digitalized using the ADCs with two samples per symbol, and then the DSP is further applied. Adaptive filtering is a subject of extensive research and many results have been reported [12]. Essentially, an equalizer of butterfly structure is preferred since adaptive algorithms must be applied to recover the polarization-division multiplexed (PDM) signals. The butterfly structured equalizer is shown in Figure 1b. Here we will first focus on the physical background and discuss the principle of coherent DSP for recovering the transmitted information. The details of algorithms will be explored after that.

Why can the electrical field of the detected signal be recovered, and what are the physical reasons behind this? The answer lies in the model of the optical fiber transmission system. In order to simplify the discussion, we firstly consider the SMF system and linear impairments, which can be solved based on linear fiber optics.

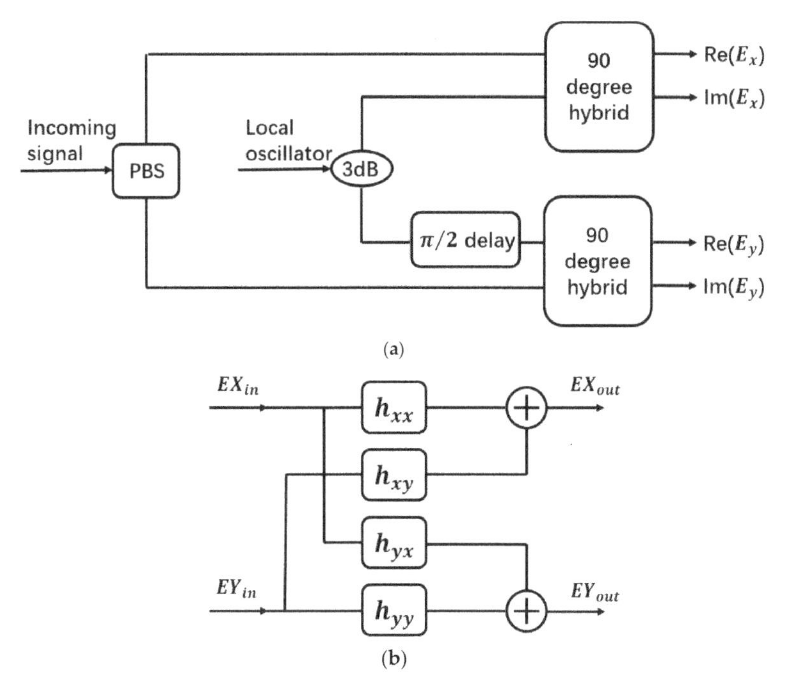

Figure 1. (**a**) Schematic of a typical coherent receiver; (**b**) Equalizer with butterfly structure.

The SMF supports the propagation of two polarized lightwaves. Optical waves usually do not remain in the principle orientation when they are propagating through the optical fiber. In the long-haul SMF system, the polarization modes are strongly coupled. Signal propagation in each section can be modeled as a 2×2 matrix. When the fiber is longer than the correlation length, it can be modeled as a concatenation of multiple sections with independent characteristics as shown in Figure 2. The overall transmission matrix H is the product of all independent matrices. No matter what the orientation of polarization beam splitters is, signals in two polarizations (EX_{out} and EY_{out}) cannot be distinguished due to the serious coupling between two degenerate modes. Therefore, the necessary condition for recovering the transmitted signals is that the transmission matrix has to be reversible (or unitary for the best case). The output electrical field can then be linked to the input electrical field by the matrix. The polarization state variation during transmission can be deduced through a unitary Jones matrix J. Regardless of the polarization dependent loss or signal attenuation (fully compensated through amplification), the overall transmission matrix H is definitely unitary and of course reversible. Compared with the CD, which can be regarded as constant, the channel transfer matrix varies with time, due to the rapid and random polarization coupling, so the adaptive DSP scheme has to be utilized. The equalizer with butterfly structure has been widely utilized. This type of equalizer can demultiplex the PDM signals with significant crosstalk and can also equalize linear impairments well. The reverse transfer matrix H^{-1} is comprised of four elements. Each element includes a tap-weight vector based finite impulse response (FIR) filter. The length of adaptive digital filter should be equal to, or a bit longer than, the impulse response spread of the distorted signal. A variety of polarization demultiplexing algorithms have been proposed, such as constant modulus algorithm (CMA), least mean square (LMS), recursive least square (RLS), radius directed equalization (RDE) and so on [13,14]. The details of the CMA algorithm are described below, and the tap weights vector of the equalizer is adapted by the following:

$$\begin{aligned} h_{xx} &= h_{xx} + \mu\varepsilon_{CMA,X} EX_{out} \cdot EX_{in}^{*} \\ h_{xy} &= h_{xy} + \mu\varepsilon_{CMA,X} EX_{out} \cdot EY_{in}^{*} \\ h_{yx} &= h_{yx} + \mu\varepsilon_{CMA,Y} EY_{out} \cdot EX_{in}^{*} \\ h_{yy} &= h_{yy} + \mu\varepsilon_{CMA,Y} EY_{out} \cdot EY_{in}^{*} \end{aligned} \quad (1)$$

$$\begin{aligned} \varepsilon_{CMA,X} &= 1 - |EX_{out}|^2 \\ \varepsilon_{CMA,Y} &= 1 - |EY_{out}|^2 \end{aligned} \quad (2)$$

where μ is the iteration factor, ε is the result of the error function. The "·" denotes the vector dot product. E^* denotes for the complex conjugate form. All tap weights are typically set to zero initially, except for the central taps of h_{xx} and h_{yy} which are set to unity. The sample rate is twice the symbol rate while the filter tap weights are updated every two samples. The equalization outputs for the two polarizations are:

$$\begin{aligned} EX_{out} &= h_{xx} \cdot EX_{in} + h_{xy} \cdot EY_{in} \\ EY_{out} &= h_{yx} \cdot EX_{in} + h_{yy} \cdot EY_{in} \end{aligned} \quad (3)$$

Another widely employed algorithm is called LMS. LMS is a type of the stochastic gradient algorithms. Tap weights are determined by the update scheme as follows.

$$\begin{aligned} h_{xx} &= h_{xx} + \mu\varepsilon_{LMS,X} EX_{in}^{*} \\ h_{xy} &= h_{xy} + \mu\varepsilon_{LMS,X} EY_{in}^{*} \\ h_{yx} &= h_{yx} + \mu\varepsilon_{LMS,Y} EX_{in}^{*} \\ h_{yy} &= h_{yy} + \mu\varepsilon_{LMS,Y} EY_{in}^{*} \end{aligned} \quad (4)$$

$$\begin{aligned} \varepsilon_{LMS,X} &= R_X - EX_{out} \\ \varepsilon_{LMS,Y} &= R_Y - EY_{out} \end{aligned} \quad (5)$$

$$\begin{aligned} R_X &= \exp(j\theta_x) d_x \\ R_Y &= \exp(j\theta_y) d_y \end{aligned} \quad (6)$$

where EX_{in}^{*}, EY_{in}^{*} represent the complex conjugate of the sampled input signal vectors and EX_{out}, EY_{out} is the output of the equalizer. Its equalization error is defined by the difference between the reference signal R and the output signal, which includes both the amplitude and the phase information. R is the reference signal, d_x, d_y are training signals or decision signals, θ_x, θ_y are the estimated symbol phase after frequency offset compensation and carrier phase estimation (CPE). At the beginning, the equalizer works in the training symbol (TS) mode. Once the equalizer has converged, it moves into a decision-directed (DD) mode. This working scheme is defined as DD-LMS.

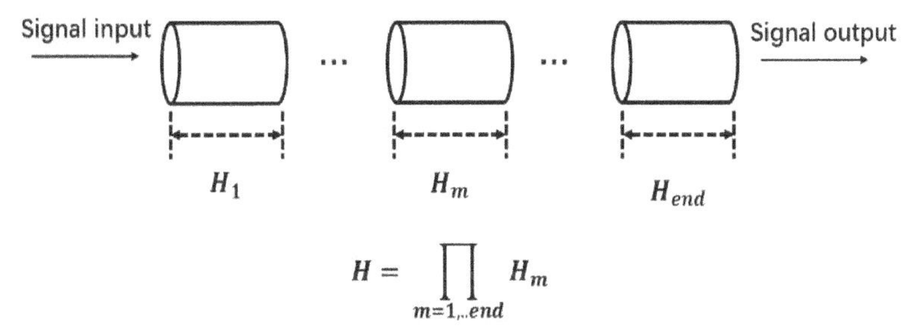

Figure 2. Model of the linear fiber transmission link.

Before the final stage of restoring the data and applying forward error correction (FEC), specific algorithms are employed to execute the frequency offset compensation and CPE [15–21]. In theory, coherent detection requires the frequency and the phase of the LO wave to be exactly the same as those of the signal carrier. However, due to the influence of fabrication imperfection of optical devices and environment variation, the frequencies of the transmitter and the LO lasers would not be completely consistent. Besides, the linewidth of the lasers will also introduce the corresponding phase noise, which is always considered as a Wiener process. The large amount of additional phase noise is quite harmful to the phase modulated signal. The purpose of the carrier recovery algorithm in the coherent receiver is to remove the impairments of carrier frequency offset and phase noise by processing a discrete data sample sequence. The principle of the frequency offset estimation is shown in Figure 3. In the case of considering the symbol phase only, it is assumed that the sampling value of the k_{th} symbol received is:

$$S(k) = \exp\{j(\theta_s(k) + \Delta\omega kT + \theta_L(k) + \theta_{ASE}(k))\} \tag{7}$$

where $\theta_s(k)$ represents the modulated phase, $\Delta\omega kT$ is additional phase caused by frequency offset, $\theta_L(k)$ is the phase noise from the laser linewidth, $\theta_{ASE}(k)$ is related to the amplified spontaneous emission (ASE) noise, T is the symbol period. In high-speed optical transmission systems, θ_L varies slowly relative to the symbol rate. By calculating the phase difference between the adjacent symbols, the θ_L can be removed.

$$S(k)S^*(k-1) = \exp\{j(\Delta\theta_s + \Delta\omega T + \theta_{ASE})\} \tag{8}$$

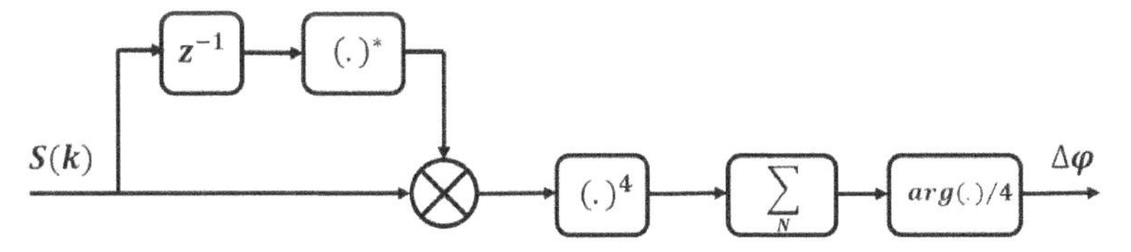

Figure 3. Block diagram of the frequency estimator.

Take the quadrature-phase-shift-keying (QPSK) signal as an example. The QPSK signal has two types of modulated phase which are $(0, \frac{\pi}{2}, \pi, \frac{3\pi}{2})$ and $(\frac{\pi}{4}, \frac{3\pi}{4}, \frac{5\pi}{4}, \frac{7\pi}{4})$, but the $\Delta\theta_s$ will always be $(0, \frac{\pi}{2}, \pi, \frac{3\pi}{2})$. The modulated phase will disappear by a power of four calculation:

$$(S(k)S^*(k-1))^4 = \exp\{j(4\Delta\omega T + 4\theta_{ASE})\} \tag{9}$$

In high-speed optical transmission systems, the frequency shift is also a slow variation process relative to the symbol rate, so that the frequency offset corresponding to multiple consecutive symbols can be regarded as the same. In this case, a series of consecutive symbols in a block can be processed together to estimate the frequency offset. The effect of θ_{ASE} is quite small after the average operation and can be neglected at the optical SNR scenario. Then the frequency offset is obtained through:

$$\sum_{N} (S(k)S^{*}(k-1))^{4}/N = \exp\{j(4\Delta\hat{\omega}T)\} \tag{10}$$

$$\arg\{\sum_{N} (S(k)S^{*}(k-1))^{4}/N\}/4 = \Delta\hat{\omega}T \tag{11}$$

After the frequency offset estimation and compensation, carrier phase recovery is used to remove the phase noise component θ_L and $\Delta\omega' KT$ caused by the linewidth and residual frequency offset of the lasers at the transceiver $\Delta\omega' = \Delta\omega - \Delta\hat{\omega}$. Viterbi–Viterbi phase estimation algorithm is a commonly used feed-forward digital CPE algorithm [22]. Figure 4 shows the principle diagram of the Viterbi–Viterbi CPE. The processing flow of the algorithm is similar to the aforementioned frequency offset estimation algorithm. The main steps are shown as follows:

$$\begin{aligned} S'(k) &= \exp\{j(\theta_s(k) + \theta_L(k) + \Delta\omega' kT + \theta_{ASE})\} \\ &= \exp\{j(\theta_s(k) + \theta_L{}'(k) + \theta_{ASE})\} \end{aligned} \tag{12}$$

where $\theta_L{}'(k)$ is the carrier phase noise of the symbol k, $\theta_L{}'(k) = \theta_L(k) + \Delta\omega' kT$

$$(S'(k))^{4} = \exp\{j(4\theta_L{}'(k) + 4\theta_{ASE})\} \tag{13}$$

θ_L caused by the laser linewidth is the main part of the $\theta_L{}'$. It changes slowly, in contrast to the high-speed symbol stream and can basically be regarded as stable in a continuous symbol block. M symbols are considered to eliminate the effect of θ_{ASE} to increase the accuracy of estimation by means of averaging symbols.

$$\sum_{M} (S'(k))^{4}/M = \exp\{j(4\hat{\theta}_L{}')\} \tag{14}$$

$$\arg(\sum_{M} (S'(k))^{4}/M)/4 = \hat{\theta}_L{}' \tag{15}$$

Equation (15) is suitable for the $(0, \frac{\pi}{2}, \pi, \frac{3\pi}{2})$ QPSK modulated signal. The residual value of modulated phase π will exist in the symbolic phase after the power of four operations if the $(\frac{\pi}{4}, \frac{3\pi}{4}, \frac{5\pi}{4}, \frac{7\pi}{4})$ modulation is applied. Then Equation (15) will be expressed as:

$$(\arg(\sum_{M} (S'(k))^{4}/M) - \pi)/4 = \hat{\theta}_L{}' \tag{16}$$

The carrier phase error calculated by M symbols will be shared with M symbols for phase compensation:

$$S''(k) = S'(k)\exp(-j\hat{\theta}_L{}') \tag{17}$$

Some typical achievements have been reported recently. Savory et al. demonstrated a 10 Gbaud PDM-QPSK experiment over 6400 km with DSP equalization [23]. The experimental results of equalized constellation diagrams of the two polarizations are shown in Figure 5. Zhou et al. proposed and demonstrated 400 Gbit/s experiments on a 50 GHz grid and successfully achieved 1200 km and 4000 km transmission [10]. Winzer et al. reported a 56 Gbaud, 224 Gbit/s PDM-QPSK 2500 km transmission experiment assisted by coherent detection with all linear impairments compensation. The experiment results of BER and transmission performance are shown in Figure 6a,b [24]. The authors of [25] developed 10-channel 28 Gbaud PDM 16-ary quadrature amplitude modulation (PDM-16QAM) signals transmitted over 1200 km.

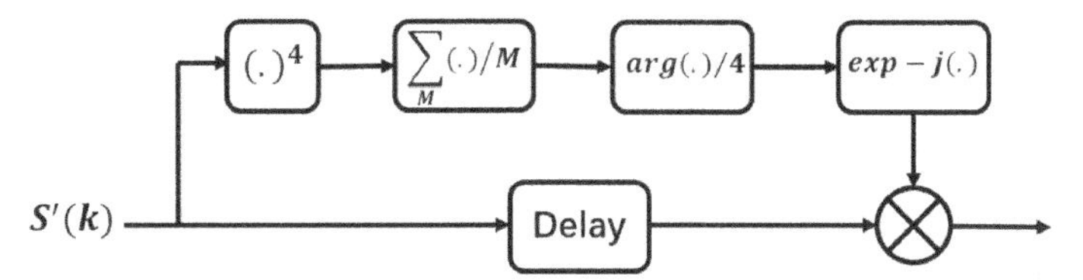

Figure 4. Block diagram of carrier phase estimator.

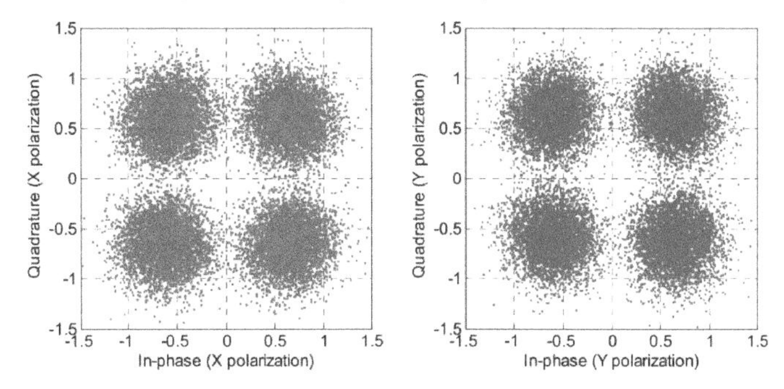

Figure 5. Recovered constellation diagrams for the two polarizations after 6400 km transmission with an estimated BER = 2.4×10^{-3}.

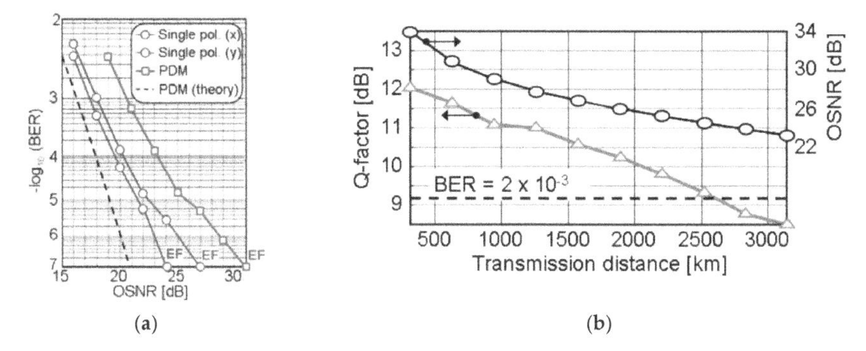

Figure 6. (**a**) B2B BER performance at 56 Gbaud: single polarization 112 Gbit/s quadrature-phase-shift-keying (QPSK) and 224 Gbit/s polarization-division multiplexed (PDM)-QPSK; (**b**) Q-factor and optical signal-to-noise ratio (OSNR) vs. transmission distance.

3. Nonlinearity Compensation

Essentially, optical signals will suffer linear and nonlinear impairments during propagation along optical fibers, which limit the transmission capacity. When WDM is introduced to increase the capacity of optical fibers, the nonlinearity and dispersion can significantly affect the signal quality. Therefore, mitigating or compensating these impairments is an important part of optical fiber communication research. At present, the method of compensating linear impairments (including CD and PMD) is almost mature, so the nonlinear impairment becomes the limiting factor of increasing optical fiber system capacity. Enlarging the number of WDM channels or reducing the channel spacing will increase the nonlinear impairment. In addition, in order to use the higher-order modulation format to improve the SE and increase the transmission distance, it is necessary to improve the OSNR. Increasing the OSNR to a given level requires the increase of signal power, which in turn, results in more serious nonlinear impairments. Therefore, the reduction or compensation of nonlinear impairments can significantly increase the capacity of optical fiber channels.

Recent developments in coherent optical communication have enabled the complex electric field of signals to be obtained in the digital domain. Thus, it is possible to compensate for nonlinear impairments in the signal waveforms propagating along the optical fiber. Furthermore, as long as adjacent channels are received, compensation for nonlinear impairments among channels, such as cross-phase modulation (XPM) and four-wave mixing (FWM), is also possible. Among the proposed nonlinearity compensation technologies, the digital backward propagation (DBP) has proved to be the most promising approach. The recent proposed investigation has demonstrated that the nonlinear Schrödinger equation (NLSE) can be solved in the digital domain to compensate all deterministic fiber transmission impairments in a PDM WDM environment. The DBP method is based on solving the nonlinear propagation equation along the backward direction. However, the computational load of DBP is heavy. It is still very important to investigate computational efficiently nonlinearity compensation algorithms for dispersion-unmanaged coherent transmission systems. Li et al. have demonstrated approaches to reduce computational load including implementing DBP using coupled NLSEs [26,27], XPM walk-off factorization [28] and folded DBP for dispersion-managed systems, which have the potential to reduce computational load to the same order of magnitude for dispersion compensation for a dispersion unmanaged system [29].

The schematic of DBP method is shown in Figure 7. A is the complex electric field of the received signal, $\hat{D}$ and $\hat{N}$ are the linear and the nonlinear operators, respectively. The signal is distorted in the real transmission fiber and is compensated through the virtual fiber. The received signal is processed by a digital model with opposite propagation parameters after it is detected at the receiver after transmission. Backward propagated NLSE can be expressed as [1]:

$$\frac{\partial A}{\partial(-z)} = (\hat{D} + \hat{N})A \tag{18}$$

$$\hat{D} = -j\frac{\beta_2}{2}\frac{\partial^2}{\partial t^2} + \frac{\beta_3}{6}\frac{\partial^3}{\partial t^3} - \frac{\alpha}{2} \tag{19}$$

$$\hat{N} = j\gamma|A|^2 \tag{20}$$

where α, β_2, β_3 and γ are the attenuation, first- and second-order group velocity dispersion and fiber nonlinear coefficients, respectively.

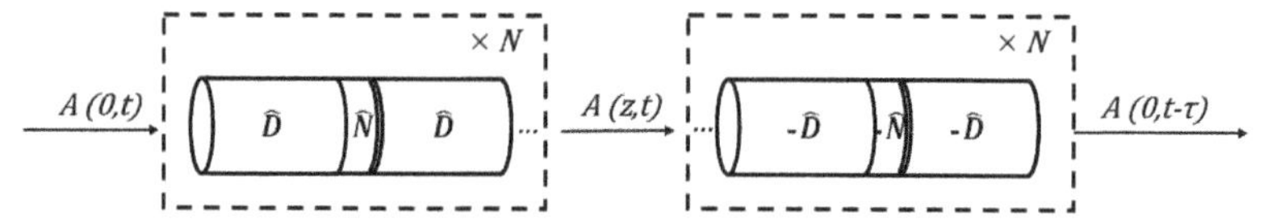

Figure 7. Schematic of digital back propagation (DBP) using the symmetric split-step method (SSM).

The performance of DBP algorithm mainly depends on the accurate estimation of NLSE propagation parameters. NLSE is usually solved by split-step Fourier method (SSFM). The optical fiber is divided into N segments, and each section has to be small enough to guarantee the optical power not to change much within the segment, since the nonlinear effects are power dependent. Another role is to ensure that the accumulation of dispersion and nonlinear effects is not so strong so as to cause serious interactions. Then the dispersion and the nonlinear effects can be considered independently in each small section, and can be executed successively (as an approximation to the simultaneous effect). In SSFM, the linear part and the nonlinear part are solved separately. The linear part is solved in frequency domain while the nonlinear part is solved in the time domain. Using backward propagated NLSE equation, all deterministic impairments (linear and nonlinear) in optical fiber transmission can be compensated. Compensation can be performed at the transmitter before signal transmission (pre-compensation) or at the receiver after signal detection (post-compensation). In most cases, DBP is implemented in conjunction with coherent receivers as a post-compensation. In the absence of noise in the transmission link, the two DBP schemes are equivalent. Since the backward propagation operator acts on the complex envelope of the electrical field, the DBP algorithm can be applied to any modulation format in principle. The performance of DBP is limited by ASE noise, because it is a stochastic noise source and cannot be compensated. DBP can only compensate for deterministic signal distortions.

The team from the University of Central Florida has studied the issue of nonlinearity compensation in WDM systems [26–28,30,31]. Nonlinear effects in optical fibers include self-phase modulation (SPM), XPM and FWM. Three nonlinear effects in optical fibers are separated using the coupling equation, and the system performance can be optimized by neglecting the items that have little influence on equalization results, but that require significant computation load. Compared with SPM and XPM, the computation load of FWM effect is much heavier. However, FWM needs phase matching to have a significant impact and the FWM effect can be neglected in non-zero dispersion shift optical fiber transmission system because it is far from the phase matching point. In this type of system, FWM items can be omitted and only the effects of SPM and XPM need to be compensated and the computational burden is reduced [26]. The simulation results are shown in Figures 8–10. Figures 9 and 10 show that XPM and FWM compensation produces the same Q value in 100 GHz channel spacing, and the influence of FWM is negligible. Larger channel spacing leads to a higher phase mismatch, and the FWM will contribute even less.

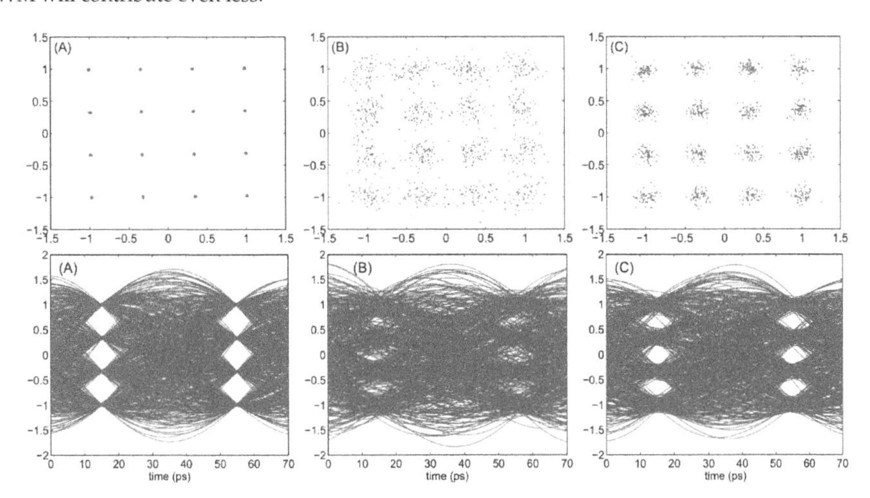

Figure 8. Constellation and eye diagrams. (**A**) B2B; (**B**) After chromatic dispersion (CD) compensation; (**C**) After cross-phase modulation (XPM) compensation.

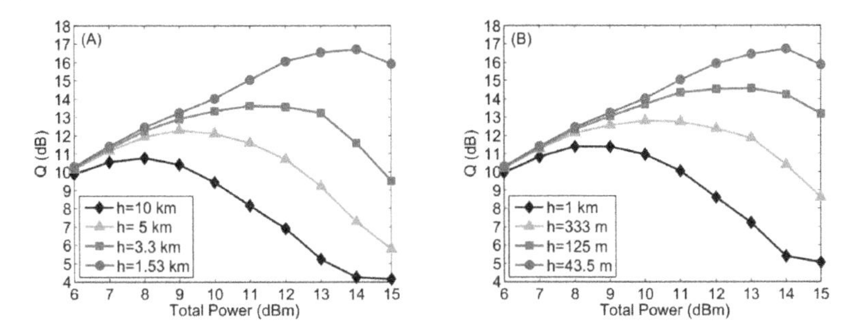

Figure 9. Received Q factor for channel spacing is 100 GHz with (**A**) XPM compensation and (**B**) XPM + four-wave mixing (FWM) compensation.

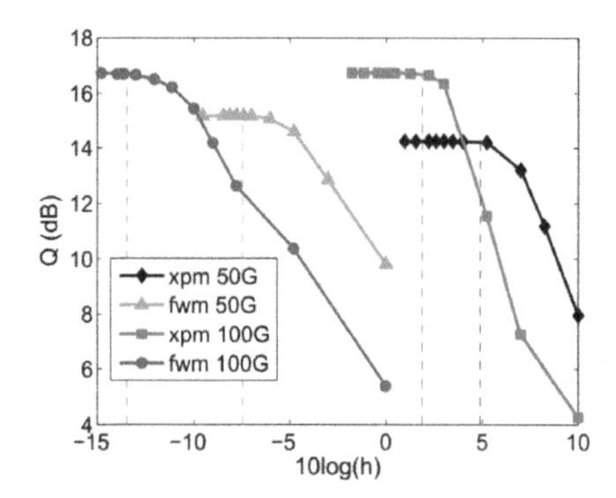

Figure 10. Q factor and step size for XPM and FWM compensation within the 50 GHz and 100 GHz grids.

But in the low dispersion region, FWM with phase matching condition will affect the performance of the transmitted signal significantly. There is a trade-off between the computational cost and the compensation effect for the FWM mitigation. The influence of the nearest channel and the second nearest channel has to be considered, while the influence of the far channel could be neglected. Using this approximation, the Q value of the observed channel can be increased with a small amount of computation [27]. The interchannel nonlinear effect XPM can be reduced by the walk-off effect caused by dispersion. The step size of SSFM can be greatly increased and the computation cost can be reduced by separating walk-off effect [28]. As shown in Figure 11, the performance of DSP improves dramatically for the XPM compensation. The step size requirements for XPM compensation using the conventional approach, and the approach by the authors of [28], are shown in Figure 12. It indicates that the step size can be increased substantially based on the SSFM. For the PDM WDM system, it can be treated by solving coupled Manakov equations, ignoring the FWM item with high computational requirements, and only compensating the XPM effect [30,31].

DBP has been experimentally studied in a series of recent works from University College London. Makovejs et al. have studied the performance improvement of single channel PDM-16QAM 112 Gbit/s system after nonlinear compensation. It is found that the maximum transmission distance can be increased from 1440 km (EDC only) to 2400 km (DBP) [32]. Savory et al. experimentally studied 112 Gbit/s non-return to zero (NRZ) PDM-QPSK WDM system. It was found that using intrachannel nonlinear compensation, the maximum transmission distance of a single channel system was increased by 46% while that of a WDM system with 100 GHz channel spacing was increased by 23% respectively [33]. Millar et al. studied 10.7 Gbaud PDM-QPSK and PDM-16QAM systems and reported that the transmission reaches have been extended to 7780 km and 1600 km, respectively [34]. Behrens et al. studied a 224 Gbit/s PDM-16QAM system and proved that the maximum transmission distance can be increased by 18–25% with the DBP implemented [35].

A 112 Gbit/s PDM-QPSK system has been studied and proposed by University College Cork in 2011. The simplification method revealed the correlation effect of adjacent symbols and realized a remarkable reduction of computation cost [36]. Similarly, the use of a correlated backpropagation (CBP) for nonlinearity compensation has been reported in a 112 Gbit/s PDM-QPSK system [37], and the step size of DBP can be increased by 70%. Dou et al. have proposed a low-complexity predistortion method for intrachannel nonlinearity compensation and have experimentally verified this approach in a 43 Gbit/s PDM-QPSK coherent system [38]. Other groups also made impressive contributions to this area, and more details can be found in Refs. [39–42].

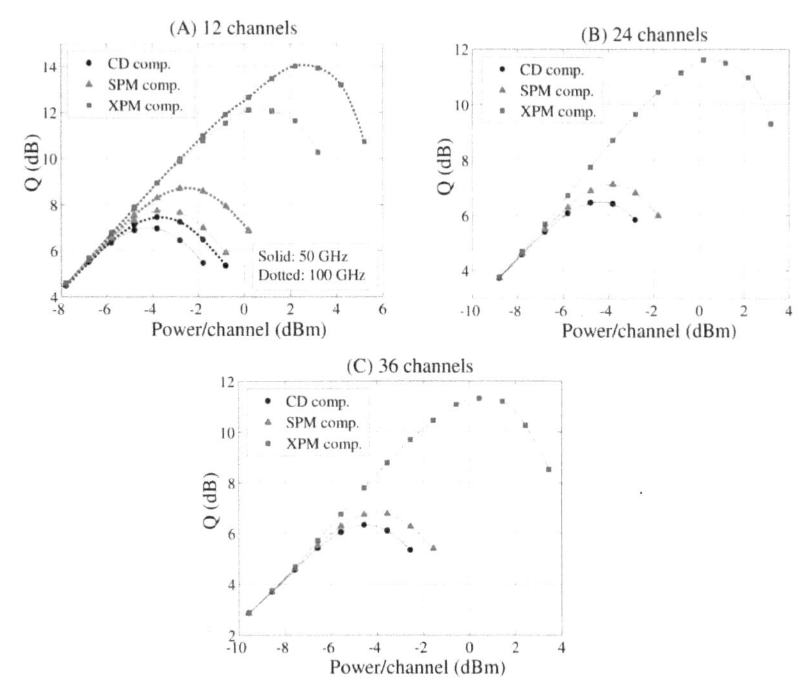

Figure 11. Performance results for: (**A**) 12 channels, (**B**) 24 channels and (**C**) 36 channels respectively.

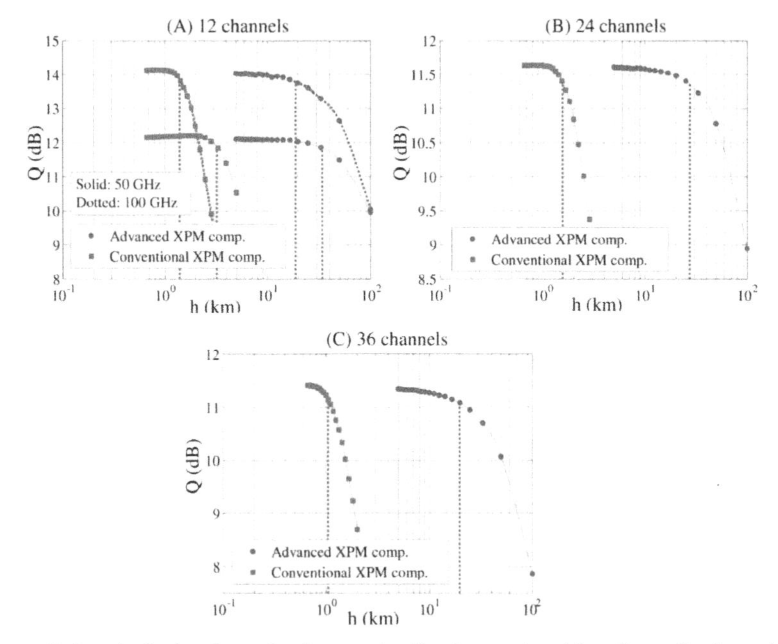

Figure 12. Step size for the advanced and conventional implementation of the split-step Fourier method (SSFM): (**A**) 12 channels, (**B**) 24 channels and (**C**) 36 channels respectively.

4. SDM and Frequency Domain Equalization (FDE)

SDM can increase the system capacity and exceed the nonlinear Shannon limit per fiber. Many efforts have been made in this area and important results have been presented. The maximum number of multiplexing has reached 72 and the capacity has reached beyond 10 Pbit/s. The authors of [43] reported a 10 × 112 Gb/s PDM-QPSK transmission in FMF over 5032 km as quasi-signal mode (QSM) application, after which the reach is extended to 7326 km [44]. Zhao et al. deduced the theoretical analysis result about the minimum number of spans required for a 3000 km QSM system [45]. More demonstrations such as transmission distance reach 14,350 km by using 12 core fiber were also reported [46]. E. Ip et al. demonstrated a WDM and MDM transmission with 146 channels and 3 linear polarized (LP) modes over 500 km. The experimental setup is shown in Figure 13 [47]. Another experiment by Ryf et al. demonstrated a 12 × 12 MIMO transmission over 130 km FMF with 8 wavelengths [48]. Fontaine et al. demonstrated a 30 × 30 MIMO transmission over 15 spatial modes [49] and Ryf et al. demonstrated a 72 × 72 MIMO DSP experimental result which is the current record of multiplexing numbers [50]. The SDM system capacity reached 1 Pbit/s in 2013 [51], and exceeded 2 Pbit/s two years later [52]. In 2017, an over 10 Pbit/s SDM transmission was demonstrated by Soma et al. which made a new milestone to fiber transmission field [53].

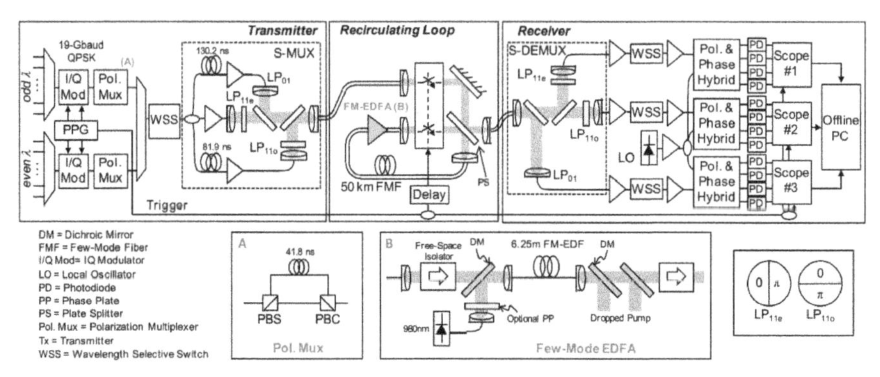

Figure 13. Experiment setup for a few-mode fiber (FMF) loop experiment.

The mode coupling and core coupling cause serious signal distortions so the MIMO DSP also needed to be applied to recover the information on each mode or fiber core. In the SDM system, the inverse transmission matrix is extended from 2×2 to $D \times D$ dimension where D is the overall coupling degree of freedom in the SDM system. The computational complexity of FIR filters increases with the number of delay taps. The transmitted signal suffers a large amount of DMGD and will continuously accumulate with the transmission distance. Large DMGD requires the digital filter with a greater number of taps, so the computational complexity will also increase dramatically. To reduce the complexity of the equalization algorithm, FDE is applied by employing discrete Fourier transform (DFT) and complex multiplications in a block-by-block way instead of convolutions in the time-domain filters [54,55]. Theoretically, the change rate of channel parameters should be much slower than the packet rate to ensure the equalization performance. Optical fiber communication systems are generally slow-varying so this condition can be easily satisfied.

The evaluation criterion for the equalization algorithm is the number of complex multiplications per symbol per mode. The computational cost of TDE scales linearly with the number of taps, while the FDE scales logarithmically with the filter length. The complexity excluding carrier recovery for these two types of algorithms can be expressed as:

$$C_{TDE} = 3m\Delta\tau LR_s \tag{21}$$

$$C_{FDE} = (4 + 4m)\log_2(2\Delta\tau L R_s) + 8m \tag{22}$$

where $\Delta\tau$ is the DMGD of the fiber, L is the link distance, R_s is the symbol rate, m can be considered as the number of degree of freedoms utilized for transmission. The algorithmic complexities of both FDE and TDE as a function of filter length are shown in Figure 14. More detailed explanations can be found in Ref. [56].

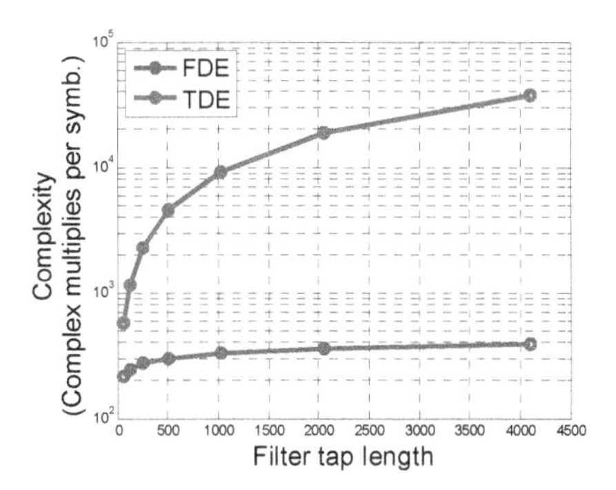

Figure 14. Complex multiplies per symbol vs. filter tap length.

As mentioned above, the computational complexity is determined by the degree of freedom and the differential mode group delay (DMGD). It is necessary to study and reduce the computational complexity of SDM applications. Several works have been reported to address this issue [57–60]. Compared with the TDE method, the FDE method can significantly reduce the computational complexity while maintaining the same performance. Some experimental results have been presented for SDM transmission, achieved by FDE. We have experimentally demonstrated an RLS FDE approach for MDM with good performance. The block diagram of the proposed algorithm and experimental results are shown in Figures 15 and 16 [60].

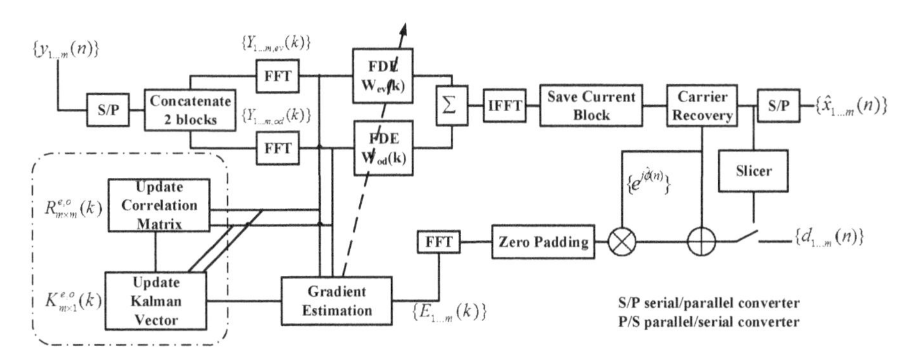

Figure 15. Block diagram of recursive least square (RLS) frequency domain equalization (FDE) for m-dimension system operating on two samples per symbol.

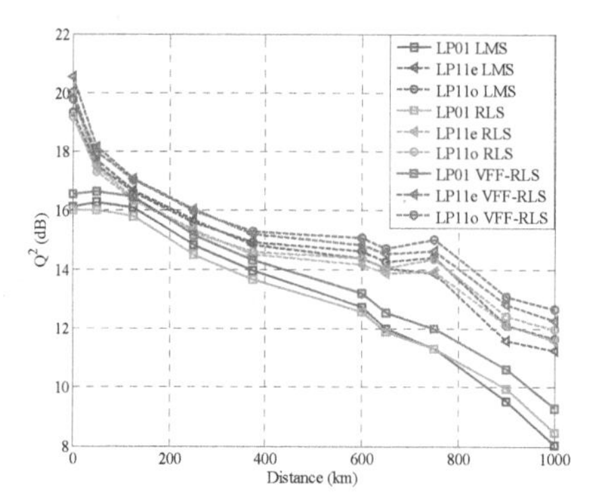

Figure 16. Comparison between conventional least mean square (LMS) and VFF-RLS/RLS: Q^2 vs. Distance.

5. AI for Optical Fiber Communication

Optical communication networks and systems have started to use artificial intelligence (AI) technologies to improve the performance, from equipment to control and management. Machine learning (ML), as a main branch of AI, provides many powerful models for optical communication. One of the main applications of ML technology is to perform optimal classification [61–63]. Demodulation of signals in point-to-point optical communication systems is crucial for achieving high-capacity transmission performance and is normally affected by nonlinear impairments and their interactions along the propagation. Classic demodulation techniques (e.g., maximum likelihood) cannot cope with high dimensional nonlinearities and channel dynamicity. Therefore, recent advances in ML and deep learning (ANNs, DNNs and CNNs) gives the potential to solve these challenges. The optimal classification can thus be directly applied for an optimal symbol detection in optical communication systems. For a memoryless nonlinearity, such as nonlinear phase noise, I/Q modulator and drive electronics nonlinearity, Euclidean distance measurement results in that the linear decision boundary is no longer optimal. For these special cases, ML techniques can be used to obtain the optimal symbol detection and decision boundary. Then DSP algorithm is used to detect the optimal symbols. ML has also been applied in many other applications such as optical performance monitoring, parameters estimation and nonlinearity mitigation [64–68]. A typical block diagram of ML for symbol classification in optical communication systems is illustrated in Figure 17.

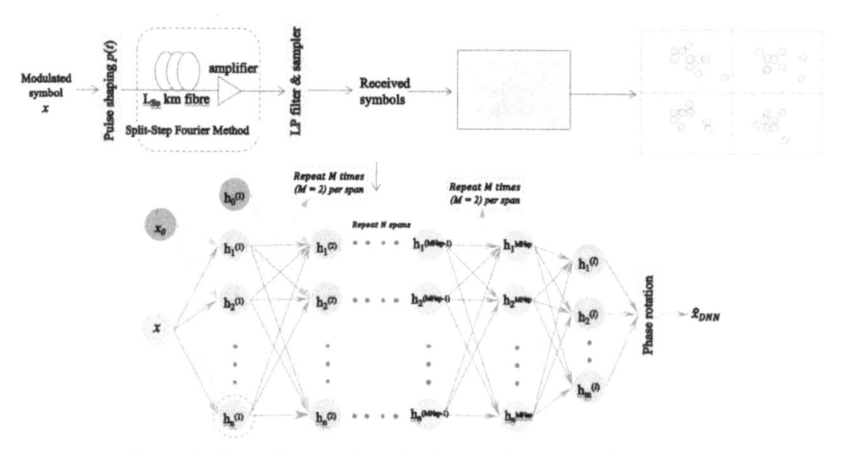

Figure 17. Block diagram of machine learning for symbol classification.

An example is implemented here to show the performance of DNN on mitigating the inter-carrier interference (ICI) in the non-orthogonal multicarrier optical fiber transmission systems with sub-carrier spacing below the symbol rate. The spectrally efficient frequency division multiplexing (SEFDM) technique is employed in such non-orthogonal multicarrier transmission systems [69,70]. The SEFDM transmission occupies a sub-carrier spacing below the symbol rate at the expense of a loss of orthogonality to enhance the achievable spectral efficiency. The challenge in the SEFDM system is the inter-carrier interference (ICI) during the bandwidth compression [71]. As illustrated in Figure 18, five types of neural networks are designed for suppressing the ICI of non-orthogonal signals [72], including no connection-neural network (N-NN), partial connection-neural network (P-NN), hybrid1-neural network (H1-NN), hybrid2-neural network (H2-NN), full connection-neural network (F-NN). These NN architectures represent different connectivity between the input, output and hidden cells in the neural networks. The complexity also depends on the NN architecture. The N-NN has the lowest computational complexity since all the input cells are independent and can be operated in parallel. The F-NN has the highest computational complexity since all neurons are connected and each connection link requires arithmetical computation.

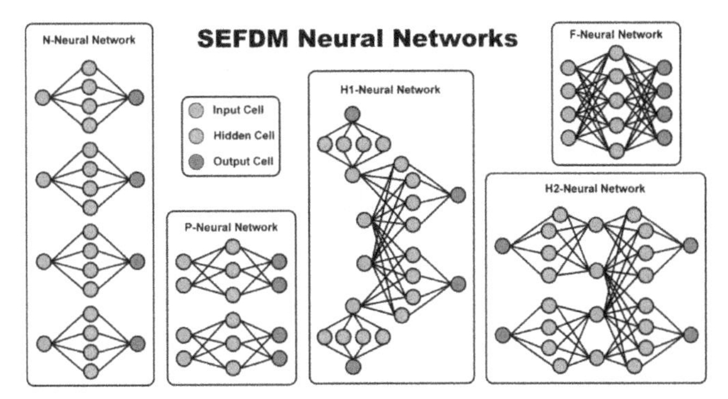

Figure 18. Neural network structures for inter-carrier interference (ICI) suppression in spectrally efficient frequency division multiplexing (SEFDM) systems.

The employed deep neural networks follow the architectures in Figure 18, and the Sigmoid function is applied as the activation function at each layer. The neural network is trained off-line using a large amount of simulated random symbols (~40,000 QPSK) in the optical SEFDM system. The number of sub-carriers in this case is set to 4 and the bandwidth compression factor is set to $\alpha = 0.8$. Please see Ref. [72] for more detailed information regarding the transmission system.

Simulation results are shown in Figure 19 to evaluate the effectiveness of NNs on SEFDM signal recovery. Their performance is compared with traditional SEFDM signal detection (hard decision). Figure 19a indicates the B2B scenario and Figure 19b indicates the scheme of 80 km transmission. Similar behaviors can be found in both scenarios. The N-DNN shows a performance close to that of the hard-decision detection, as each input cell is independently processed. Although there are sufficient neurons and hidden layers to improve the training accuracy, the independent processing architecture results in similar performance as the hard decision. The P-DNN provide a better performance since two input cells are connected and jointly trained, where more accurate interference features can be extracted from the network. The H1-DNN network shows further improved performance since the middle input cells are connected to neighboring input cells and therefore accurate interference emulation for the middle cells can be realized. As an evolved version, H2-DNN slightly outperforms the H1-DNN, where the interference to the edge input cell can be properly modelled through the neighboring neuron connection. The F-DNN provides the best performance the since the complete interference to each sub-carrier is estimated with additional neuron connections. Therefore, the complete interference information can be extracted from the SEFDM signal and an accurate model can be trained.

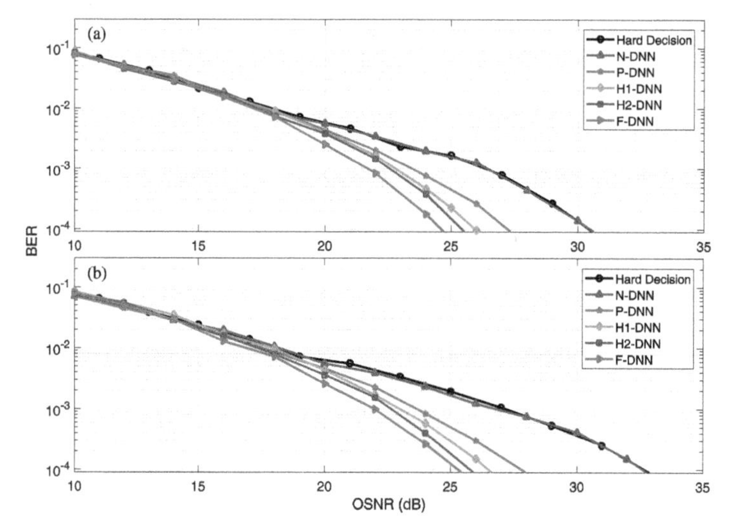

Figure 19. ICI cancellation using NNs. (**a**) B2B (back-to-back) case. (**b**) 80 km fiber transmission.

6. Conclusions

Coherent optical fiber communication has become an important field with the development of DSP techniques in both hardware and software. Fiber linear impairments can be compensated using static and adaptive filters while fiber nonlinearities can be compensated using the DBP method. SDM is considered as one of the most effective approaches to overcome the capacity crunch. The capacity of SDM system has been increased significantly with the improvement of novel DSP technologies. With the continuous reduction of the system cost and power consumption, DSP based coherent optical communication systems have been deployed widely. AI has been employed as an effective DSP tool in the coherent detection, however it still stays in the initial phase. Future research is greatly required to investigate the true capability and powerful function of ML in optical communications and other areas of optoelectronics. This paper has been focused on the applications of DSP in coherent optical systems. However, there are also applications of advanced DSP algorithms in other promising systems such as direct-detection systems which could offer lower-cost implementations of optical communications. This will be investigated in our future work.

Author Contributions: This paper was mainly written by J.Z. and T.X. revised the article. Y.L. drew Figures 1–4. J.Z. supervised the overall project.

Funding: This work was supported by the National Natural Science Foundation of China under Grants 61335005 and 61775165.

Acknowledgments: The authors would like to thank the anonymous reviewers for their valuable comments and suggestions to improve this manuscript.

Conflicts of Interest: The authors declare no conflict of interest.

References

1. Li, G. Recent advances in coherent optical communication. *Adv. Opt. Photonics* **2009**, *1*, 279–307. [CrossRef]
2. Taylor, M. Coherent detection method using DSP for demodulation of signal and subsequent equalization of propagation impairments. *IEEE Photonics Technol. Lett.* **2004**, *16*, 674–676. [CrossRef]

3. Kikuchi, K. Phase-diversity homodyne detection of multilevel optical modulation with digital carrier phase estimation. *IEEE J. Sel. Top. Quantum Electron.* **2006**, *12*, 563–570. [CrossRef]
4. Savory, S. Digital filters for coherent optical receivers. *Opt. Express* **2008**, *16*, 804–817. [CrossRef] [PubMed]
5. Ip, E.; Lau, A.; Barros, D.; Kahn, J. Coherent detection in optical fiber systems. *Opt. Express* **2008**, *16*, 753–791. [CrossRef]
6. Savory, S. Digital coherent optical receivers: Algorithms and subsystems. *IEEE J. Sel. Top. Quantum Electron.* **2010**, *16*, 1164–1179. [CrossRef]
7. Ip, E.; Kahn, J. Fiber impairment compensation using coherent detection and digital signal processing. *J. Lightwave Technol.* **2010**, *28*, 502–519. [CrossRef]
8. Yu, J.; Zhou, X. Ultra-high-capacity DWDM transmission system for 100G and beyond. *IEEE Commun. Mag.* **2010**, 56–64. [CrossRef]
9. Noe, R.; Pfau, T.; El-Darawy, M.; Hoffmann, S. Electronic polarization control algorithms for coherent optical transmission. *IEEE J. Sel. Top. Quantum Electron.* **2010**, *16*, 1193–1200. [CrossRef]
10. Zhou, X.; Nelson, L. 400G WDM transmission on the 50 GHz grid for future optical networks. *J. Lightwave Technol.* **2012**, *30*, 3779–3792. [CrossRef]
11. Han, Y.; Li, G. Coherent optical communication using polarization multiple input multiple output. *Opt. Express* **2005**, *13*, 7527–7534. [CrossRef] [PubMed]
12. Haykin, S. *Adaptive Filter Theory*, 5th ed.; Pearson Education: Hamilton, ON, Canada, 2005.
13. Fan, Y.; Chen, X.; Zhou, W.; Zhou, X.; Zhu, H. The comparison of CMA and LMS equalization algorithms in optical coherent receivers. In Proceedings of the International Conference on Wireless Communications Networking and Mobile Computing (WiCOM), Chengdu, China, 23–25 September 2010.
14. Fatadin, I.; Ives, D.; Savory, S. Blind equalization and carrier phase recovery in a 16-QAM optical coherent system. *J. Lightwave Technol.* **2009**, *27*, 3042–3049. [CrossRef]
15. Cao, Y.; Yu, S.; Shen, J.; Gu, W.; Ji, Y. Frequency estimation for optical coherent MPSK system without removing modulated data phase. *IEEE Photonics Technol. Lett.* **2010**, *22*, 691–693. [CrossRef]
16. Pfau, T.; Hoffmann, S.; Noe, R. Hardware-efficient coherent digital receiver concept with feedforward carrier recovery for M-QAM constellations. *J. Lightwave Technol.* **2009**, *27*, 989–999. [CrossRef]
17. Zhang, S.; Kam, P.; Chen, J.; Yu, C. Decision-aided maximum likelihood detection in coherent optical phase-shift-keying system. *Opt. Express* **2009**, *17*, 703–715. [CrossRef] [PubMed]
18. Fatadin, I.; Ives, D.; Savory, S. Laser linewidth tolerance for 16-QAM coherent optical systems using QPSK partitioning. *IEEE Photonics Technol. Lett.* **2010**, *22*, 631–633. [CrossRef]
19. Zhou, X. An improved feed-forward carrier recovery algorithm for coherent receivers with M-QAM modulation format. *IEEE Photonics Technol. Lett.* **2010**, *22*, 1051–1053. [CrossRef]
20. Taylor, M. Phase estimation methods for optical coherent detection using digital signal processing. *J. Lightwwave Technol.* **2009**, *27*, 901–914. [CrossRef]
21. Leven, A.; Kaneda, N.; Koc, U.; Chen, Y. Frequency estimation in intradyne reception. *IEEE Photonics Technol. Lett.* **2007**, *19*, 366–368. [CrossRef]
22. Viterbi, A.; Viterbi, A. Nonlinear estimation of PSK-modulated carrier phase with application to burst digital transmission. *IEEE Trans. Inf. Theory* **1983**, *29*, 543–551. [CrossRef]
23. Savory, S.; Gavioli, G.; Killey, R.; Bayvel, P. Electronic compensation of chromatic dispersion using a digital coherent receiver. *Opt. Express* **2007**, *15*, 2120–2126. [CrossRef] [PubMed]
24. Winzer, P.; Gnauck, A.; Raybon, G.; Schnecker, M.; Pupalaikis, P. 56-Gbaud PDM-QPSK: Coherent detection and 2500-km transmission. In Proceedings of the European Conference and Exhibition on Optical Communication (ECOC), Vienna, Austria, 20–24 September 2009.
25. Gnauck, A.; Winzer, P.; Chandrasekhar, S.; Liu, X.; Zhu, B.; Peckham, D. 10 × 224-Gb/s WDM transmission of 28-Gbaud PDM 16-QAM on a 50-GHz grid over 1200 km of fiber. In Proceedings of the Optical Fiber Communications Conference (OFC), San Diego, CA, USA, 21–25 March 2010.
26. Mateo, E.; Zhu, L.; Li, G. Impact of XPM and FWM on the digital implementation of impairment compensation for WDM transmission using backward propagation. *Opt. Express* **2008**, *16*, 16124–16137. [CrossRef] [PubMed]
27. Mateo, E.; Li, G. Compensation of interchannel nonlinearities using enhanced coupled equations for digital backward propagation. *Appl. Opt.* **2009**, *48*, F6–F10. [CrossRef] [PubMed]
28. Mateo, E.; Yaman, F.; Li, G. Efficient compensation of inter-channel nonlinear effects via digital backward propagation in WDM optical transmission. *Opt. Express* **2010**, *18*, 15144–15154. [CrossRef]

29. Zhu, L.; Li, G. Nonlinearity compensation using dispersion folded digital back propagation. *Opt. Express* **2012**, *20*, 14362–14370. [CrossRef] [PubMed]
30. Yaman, F.; Li, G. Nonlinear impairment compensation for polarization-division multiplexed WDM transmission using digital backward propagation. *IEEE Photonics J.* **2010**, *2*, 816–832. [CrossRef]
31. Mateo, E.; Zhou, X.; Li, G. Improved digital backward propagation for the compensation of inter-channel nonlinear effects in polarization-multiplexed WDM systems. *Opt. Express* **2011**, *19*, 570–583. [CrossRef]
32. Makovejs, S.; Millar, D.; Lavery, D.; Behrens, C.; Killey, R.; Savory, S.; Bayvel, P. Characterization of long-haul 112Gbit/s PDM-QAM-16 transmission with and without digital nonlinearity compensation. *Opt. Express* **2010**, *18*, 12939–12947. [CrossRef]
33. Savory, S.; Gavioli, G.; Torrengo, E.; Poggiolini, P. Impact of interchannel nonlinearities on a split-step intrachannel nonlinear equalizer. *IEEE Photonics Technol. Lett.* **2010**, *22*, 673–675. [CrossRef]
34. Millar, D.; Makovejs, S.; Behrens, C.; Hellerbrand, S.; Killey, R.; Bayvel, P.; Savory, S. Mitigation of fiber nonlinearity using a digital coherent receiver. *IEEE J. Sel. Top. Quantum Electron.* **2010**, *16*, 1217–1226. [CrossRef]
35. Behrens, C.; Makovejs, S.; Killey, R.; Savory, S.; Chen, M.; Bayvel, P. Pulse-shaping versus digital backpropagation in 224Gbit/s PDM-16QAM transmission. *Opt. Express* **2011**, *19*, 12879–12884. [CrossRef] [PubMed]
36. Rafique, D.; Mussolin, M.; Forzati, M.; Martensson, J.; Chugtai, M.; Ellis, A. Compensation of intra-channel nonlinear fiber impairments using simplified digital back-propagation algorithm. *Opt. Express* **2011**, *19*, 9453–9460. [CrossRef] [PubMed]
37. Li, L.; Tao, Z.; Dou, L.; Yan, W.; Oda, S.; Tanimura, T.; Hoshida, T.; Rasmussen, J. Implementation efficient non-linear equalizer based on correlated digital back-propagation. In Proceedings of the Optical Fiber Communications Conference (OFC), Los Angeles, CA, USA, 6–10 March 2011.
38. Dou, L.; Tao, Z.; Li, L.; Yan, W.; Tanimura, T.; Hoshida, T.; Rasmussen, J. A low complexity pre-distortion method for intra-channel nonlinearity. In Proceedings of the Optical Fiber Communications Conference (OFC), Los Angeles, CA, USA, 6–10 March 2011.
39. Tao, Z.; Dou, L.; Yan, W.; Li, L. Multiplier-free intrachannel nonlinearity compensating algorithm operating at symbol rate. *J. Lightwave Technol.* **2011**, *29*, 2570–2576. [CrossRef]
40. Lin, C.; Holtmannspoetter, M.; Asif, R.; Schmauss, B. Compensation of transmission impairments by digital backward propagation for different link designs. In Proceedings of the European Conference and Exhibition on Optical Communication (ECOC), Torino, Italy, 19–23 September 2010.
41. Asif, R.; Lin, C.; Holtmannspoetter, M.; Schmauss, B. Logarithmic step-size based digital backward propagation in N-channel 112Gbit/s/ch DP-QPSK transmission. In Proceedings of the International Conference on Transparent Optical Networks (ICTON), Stockholm, Sweden, 26–30 June 2011.
42. Du, L.; Lowery, A. Improved single channel backpropagation for intra-channel fiber nonlinearity compensation in long-haul optical communication systems. *Opt. Express* **2010**, *18*, 17075–17088. [CrossRef] [PubMed]
43. Yaman, F.; Bai, N.; Huang, Y.; Huang, M.; Zhu, B.; Wang, T.; Li, G. 10 × 112Gb/s PDM-QPSK transmission over 5032 km in few-mode fibers. *Opt. Express* **2010**, *18*, 21342–21349. [CrossRef]
44. Igarashi, K.; Tsuritani, T.; Morita, I.; Tsuchida, Y.; Maeda, K.; Tadakuma, M.; Saito, T.; Watanabe, K.; Imamura, K.; Sugizaki, R.; et al. 1.03-Exabit/skm super-Nyquist-WDM transmission over 7326-km seven-core fiber. In Proceedings of the European Conference and Exhibition on Optical Communication (ECOC), London, UK, 22–26 September 2013.
45. Zhao, J.; Kim, I.; Vassilieva, O.; Ikeuchi, T.; Wang, W.; Wen, H.; Li, G. Minimizing the number of spans for terrestrial fiber-optic systems using quasi-single-mode transmission. *IEEE Photonics J.* **2018**, *10*, 7200110. [CrossRef]
46. Turukhin, A.; Sinkin, O.; Batshon, H.; Zhang, H.; Sun, Y.; Mazurczyk, M.; Davidson, C.; Cai, J.; Bolshtyansky, M.; Foursa, D.; et al. 105.1 Tb/s power-efficient transmission over 14,350 km using a 12-core fiber. In Proceedings of the Optical Fiber Communications Conference (OFC), Anaheim, CA, USA, 20–24 March 2016.

47. Ip, E.; Li, M.; Bennett, K.; Huang, Y.; Tanaka, A.; Korolev, A.; Koreshkov, K.; Wood, W.; Mateo, E.; Hu, J.; et al. 146λ×6×19-Gbaud wavelength- and mode-division multiplexed transmission over 10 × 50-km spans of few-mode fiber with a gain-equalized few-mode EDFA. In Proceedings of the Optical Fiber Communications Conference (OFC), Anaheim, CA, USA, 17–21 March 2013.
48. Ryf, R.; Fontaine, N.; Mestre, M.; Randel, S.; Palou, X.; Bolle, C.; Gnauck, A.; Chandrasekhar, S.; Liu, X.; Guan, B.; et al. 12 × 12 MIMO Transmission over 130-km Few-Mode Fiber. In Proceedings of the Frontiers in Optics, Orlando, FL, USA, 6–10 October 2013.
49. Fontaine, N.; Ryf, R.; Chen, H.; Benitez, A.; Antonio Lopez, J.; Correa, R.G.; Guan, B.; Ercan, B.; Scott, R.; Ben Yoo, S.; et al. 30 × 30 MIMO Transmission over 15 Spatial Modes. In Proceedings of the Optical Fiber Communications Conference (OFC), Los Angeles, CA, USA, 22–26 March 2015.
50. Ryf, R.; Fontaine, N.; Chen, H.; Wittek, S.; Li, J.; Alvarado-Zacarias, J.; Amezcua-Correa, R.; Antonio-Lopez, J.; Capuzzo, M.; Kopf, R.; et al. Mode-Multiplexed Transmission over 36 Spatial Modes of a Graded-Index Multimode Fiber. In Proceedings of the European Conference and Exhibition on Optical Communication (ECOC), Roma, Italy, 23–27 September 2018.
51. Takara, H.; Sano, A.; Kobayashi, T.; Kubota, H.; Kawakami, H.; Matsuura, A.; Miyamoto, Y.; Abe, Y.; Ono, H.; Shikama, K.; et al. 1.01-Pb/s (12 SDM/222 WDM/456 Gb/s) Crosstalk-managed Transmission with 91.4-b/s/Hz Aggregate Spectral Efficiency. In Proceedings of the European Conference and Exhibition on Optical Communication (ECOC), London, UK, 22–26 September 2013.
52. Puttnam, B.; Luis, R.; Klaus, W.; Sakaguchi, J.; Delgado Mendinueta, J.; Awaji, Y.; Wada, N.; Tamura, Y.; Hayashi, T.; Hirano, M.; et al. 2.15 Pb/s transmission using a 22 core homogeneous single-mode multi-core fiber and wideband optical comb. In Proceedings of the European Conference and Exhibition on Optical Communication (ECOC), Valencia, Spain, 27 September–1 October 2015.
53. Soma, D.; Wakayama, Y.; Beppu, S.; Sumita, S.; Tsuritani, T.; Hayashi, T.; Nagashima, T.; Suzuki, M.; Takahashi, H.; Igarashi, K.; et al. 10.16 Peta-bit/s Dense SDM/WDM transmission over Low-DMD 6-Mode 19-Core Fibre across C+L Band. In Proceedings of the European Conference and Exhibition on Optical Communication (ECOC), Gothenburg, Sweden, 17–21 September 2017.
54. Faruk, M.; Kikuchi, K. Adaptive frequency-domain equalization in digital coherent optical receivers. *Opt. Express* **2011**, *19*, 12789–12798. [CrossRef]
55. Kudo, R.; Kobayashi, T.; Ishihara, K.; Takatori, Y.; Sano, A.; Miyamoto, Y. Coherent optical single carrier transmission using overlap frequency domain equalization for long-haul optical systems. *J. Lightwave Technol.* **2009**, *27*, 3721–3728. [CrossRef]
56. Bai, N.; Li, G. Adaptive frequency-domain equalization for mode-division multiplexed transmission. *IEEE Photonics Technol. Lett.* **2012**, *24*, 1918–1921.
57. Zhu, C.; Tran, A.; Chen, S.; Du, L.; Anderson, T.; Lowery, A.; Skafidas, E. Improved two-stage equalization for coherent Pol-Mux QPSK and 16-QAM systems. *Opt. Express* **2012**, *20*, B141–B150. [CrossRef]
58. Bai, N.; Xia, C.; Li, G. Adaptive frequency-domain equalization for the transmission of the fundamental mode in a few-mode fiber. *Opt. Express* **2012**, *20*, 24010–24017. [CrossRef] [PubMed]
59. Arık, S.; Askarov, D.; Kahn, J. Adaptive frequency-domain equalization in mode-division multiplexing systems. *J. Lightwave Technol.* **2014**, *32*, 1841–1852. [CrossRef]
60. Yang, Z.; Zhao, J.; Bai, N.; Ip, E.; Wang, T.; Li, Z.; Li, G. Experimental demonstration of adaptive VFF-RLS-FDE for long-distance mode-division multiplexed transmission. *Opt. Express* **2018**, *26*, 18362–18367. [CrossRef] [PubMed]
61. Wang, D.; Zhang, M.; Li, Z.; Cui, Y.; Liu, J.; Yang, Y.; Wang, H. Nonlinear decision boundary created by a machine learning-based classifier to mitigate nonlinear phase noise. In Proceedings of the European Conference and Exhibition on Optical Communication (ECOC), Valencia, Spain, 27 September–1 October 2015.
62. Zibar, D.; Winther, O.; Franceschi, N.; Borkowski, R.; Caballero, A.; Arlunno, V.; Schmidt, M.; Gonzales, N.; Mao, B.; Ye, Y.; et al. Nonlinear impairment compensation using expectation maximization for dispersion managed and unmanaged PDM 16-QAM transmission. *Opt. Express* **2012**, *20*, 181–196. [CrossRef] [PubMed]
63. Li, M.; Yu, S.; Yang, J.; Chen, Z.; Han, Y.; Gu, W. Nonparameter nonlinear phase noise mitigation by using M-ary support vector machine for coherent optical systems. *IEEE Photonics J.* **2013**, *5*, 7800312. [CrossRef]
64. Argyris, A.; Bueno, J.; Fischer, I. Photonic machine learning implementation for signal recovery in optical communications. *Sci. Rep.* **2018**, *8*, 1–13. [CrossRef] [PubMed]

65. Thrane, J.; Wass, J.; Piels, M.; Diniz, J.; Jones, R.; Zibar, D. Machine learning techniques for optical performance monitoring from directly detected PDM-QAM signals. *J. Lightwave Technol.* **2017**, *35*, 868–875. [CrossRef]
66. Zibar, D.; Piels, M.; Jones, R.; Schaeffer, C. Machine learning techniques in optical communication. *J. Lightware Technol.* **2016**, *34*, 1442–1452. [CrossRef]
67. Khan, F.; Fan, Q.; Lu, C.; Lau, A. An optical communication's perspective on machine learning and its applications. *J. Lightwave Technol.* **2019**, *37*, 493–516. [CrossRef]
68. Khan, F.; Zhong, K.; Al-Arashi, W.; Yu, C.; Lu, C.; Lau, A. Modulation format identification in coherent receivers using deep machine learning. *IEEE Photonics Technol. Lett.* **2016**, *28*, 1886–1889. [CrossRef]
69. Xu, T.; Xu, T.; Bayvel, P.; Darwazeh, I. Non-orthogonal signal transmission over nonlinear optical channels. *IEEE Photonics J.* **2019**, *11*, 7203313. [CrossRef]
70. Darwazeh, I.; Xu, T.; Gui, T.; Bao, Y.; Li, Z. Optical SEFDM system; bandwidth saving using non-orthogonal sub-carriers. *IEEE Photonics Technol. Lett.* **2014**, *26*, 352–355. [CrossRef]
71. Zhao, J.; Ellis, A.D. A novel optical fast OFDM with reduced channel spacing equal to half of the symbol rate per carrier. In Proceedings of the Optical Fiber Communications Conference (OFC), San Diego, CA, USA, 21–25 March 2010.
72. Xu, T.; Xu, T.; Darwazeh, I. Deep learning for interference cancellation in non-orthogonal signal based optical communication systems. In Proceedings of the Progress in Electromagnetics Research Symposium (PIERS), Toyama, Japan, 1–4 August 2018.

Review

Recent Advances in DSP Techniques for Mode Division Multiplexing Optical Networks with MIMO Equalization: A Review

Yi Weng [1], Junyi Wang [2] and Zhongqi Pan [1,*]

[1] Department of Electrical and Computer Engineering, University of Louisiana at Lafayette, Lafayette, LA 70504, USA; wengyi.201@gmail.com
[2] Qualcomm Technologies Inc., San Diego, CA 95110, USA; wjijackcy@gmail.com
* Correspondence: zpan@louisiana.edu; Tel.: +1-337-482-5899

Received: 12 February 2019; Accepted: 7 March 2019; Published: 20 March 2019

Abstract: This paper provides a technical review regarding the latest progress on multi-input multi-output (MIMO) digital signal processing (DSP) equalization techniques for high-capacity fiber-optic communication networks. Space division multiplexing (SDM) technology was initially developed to improve the demanding capacity of optic-interconnect links through mode-division multiplexing (MDM) using few-mode fibers (FMF), or core-multiplexing exploiting multicore fibers (MCF). Primarily, adaptive MIMO filtering techniques were proposed to de-multiplex the signals upon different modes or cores, and to dynamically compensate for the differential mode group delays (DMGD) plus mode-dependent loss (MDL) via DSP. Particularly, the frequency-domain equalization (FDE) techniques suggestively lessen the algorithmic complexity, compared with time-domain equalization (TDE), while holding comparable performance, amongst which the least mean squares (LMS) and recursive least squares (RLS) algorithms are most ubiquitous and, hence, extensively premeditated. In this paper, we not only enclose the state of the art of MIMO equalizers, predominantly focusing on the advantage of implementing the space–time block-coding (STBC)-assisted MIMO technique, but we also cover the performance evaluation for different MIMO-FDE schemes of DMGD and MDL for adaptive coherent receivers. Moreover, the hardware complexity optimization for MIMO-DSP is discussed, and a joint-compensation scheme is deliberated for chromatic dispersion (CD) and DMGD, along with a number of recent experimental demonstrations using MIMO-DSP.

Keywords: digital signal processing; multi-input multi-output; mode-division multiplexing; least mean squares; frequency-domain equalization; recursive least squares; space–time block-coding; mode-dependent loss

---

1. Introduction

The space division multiplexing (SDM) systems were proposed to improve the capacity of fiber-optic transmission links, either by means of core multiplexing exploiting multicore fibers (MCF), or with mode-division multiplexing (MDM) using few-mode fibers (FMF) [1–3]. Device integration is the most likely approach to accomplish power consumption and cost reduction, through a variety of proposals for realizing parallel spatial channels, including integrating parallel transmitters, receivers, and amplifiers in the same device, as well as utilizing various multimode or multicore fibers, as shown in Figure 1 [4].

Amongst these proposals, the simplest solution is to solely apply fiber bundles to have the same behavior as N single-mode fibers, which can consist of a fiber ribbon or a multi-element fiber as displayed in Figure 1a. Such a multi-element fiber provides a smooth upgrade path where all existing components can be reused, and the power consumption and cost go up with the scale of the fiber bundles. In contrast, MCFs epitomize the next step in parallel fiber integration, where the adjacent fiber cores are enclosed in the same glass cladding, and it allows novel system architectures such as local oscillator sharing with less temperature-dependent fluctuation, as revealed in Figure 1b. It is worth mentioning that MCFs are usually designed with low inter-core cross-talk; thus, multiple-input multiple-output (MIMO) digital signal processing (DSP) is not required at the receiver. On the other hand, mode coupling is difficult to avoid in FMFs and, hence, necessitates the usage of MIMO DSP in the SDM systems, where orthogonal spatial modes within the same fiber core form the parallel channels, as exposed in Figure 1c. Furthermore, strongly coupled MCFs that behave like FMFs were also proposed with certain combined advantages, which shows there is great potential to combine MCF with few-mode cores in next-generation optical-fiber communication systems [5,6].

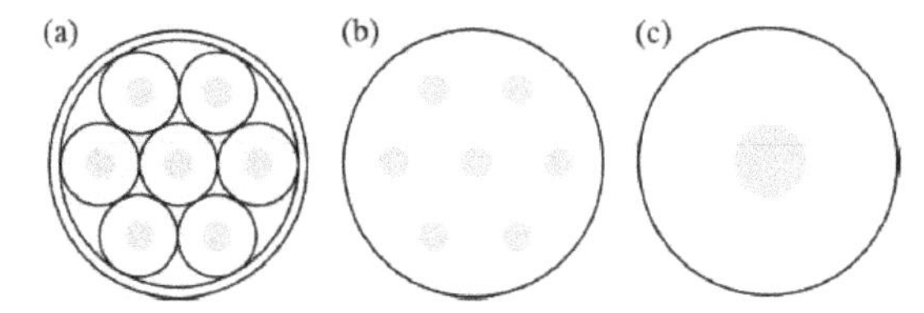

Figure 1. The typical examples of optical fibers for space division multiplexing (SDM) technology: (**a**) multi-element fiber; (**b**) multicore fiber; (**c**) multi-mode fiber [4].

Principally, the adaptive multi-input multi-output (MIMO) frequency-domain equalization (FDE) techniques were recognized to de-multiplex the signals upon diverse modes, as well as compensate for the differential mode group delay (DMGD) dynamically through digital signal processing (DSP) [7–9]. The FDE techniques can reduce the algorithmic complexity evocatively, in comparison with the time-domain equalization (TDE), while holding equivalent performance, amongst which the least mean squares (LMS) and recursive least squares (RLS) algorithms are most widely held and, hence, broadly studied [10–12]. Although the LMS approach has lower complexity, it suffers from severe performance deprivation and exhibits slower convergence, while encountering a large number of parameters to be simultaneously adapted in future SDM channels [13]. Alternatively, faster convergence can be accomplished by applying RLS with a more sophisticated algorithmic conformation at the price of higher complexity [14,15]. Hereafter, it is significant to develop a MIMO-DSP FDE algorithm with the intention of achieving an RLS-level performance at an LMS-adapted complexity for an SDM-based optical communication system.

Another bewildering deficiency of FMF-based optical fiber systems is the mode-dependent loss (MDL), which soars from inline component defectiveness and disorders the modal orthogonality, consequently degrading the overall capacity of MIMO channels [16]. With the aim of lessening the impact of MDL, old-fashioned tactics take account of mode scrambler and specialty fiber designs. These approaches were primarily disadvantaged with high cost, yet they cannot utterly eradicate the accumulated MDL in optical links [17]. Furthermore, the space–time trellis codes (STTC) were proposed to reduce the MDL, but they suffered from great complexity as well [18]. Henceforward, MDL is categorically another key challenge that needs to be triumphed for next-generation optical transmission systems.

The spatial multiplicity versus the transmission distance accomplished in recent SDM- wavelength division multiplexing (WDM) transmission experiments is plotted in Figure 2 [19], where the region with a spatial multiplicity over 30 is designated as the "DSDM region". The dots in Figure 2 represent the cases of multimode fiber (MMF), MCF, and coupled-core fibers (only three-core and six-core), as well as the latest multicore multimode fiber experiments [20].

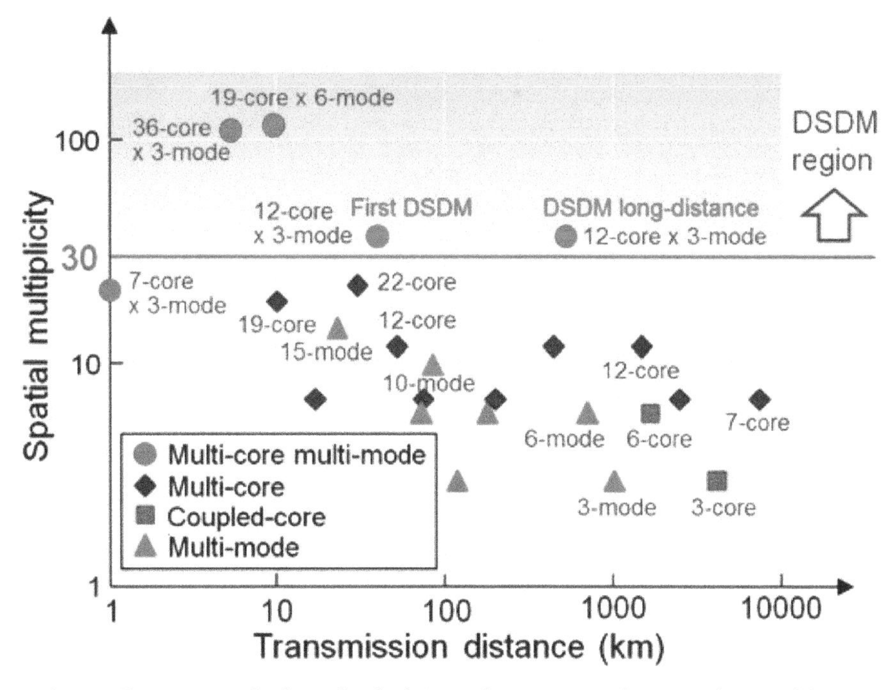

Figure 2. The recent records of spatial multiplicity vs. the transmission distance in the state-of-the-art SDM-WDM experiments [19].

The outline of this review article is sketched as follows: Section 2 describes the principles of MIMO equalizers, the LMS and RLS techniques, and the space–time block-coding (STBC) assisted RLS algorithms, while Section 3 introduces the configuration of a typical multimode coherent transmission system transmission system with FMFs for high-order modulation formats. In Section 4, the performance analysis of various adaptive filtering schemes is presented for quadrature phase shift keying (QPSK), 16 quadrature amplitude modulation (16QAM), and 64QAM, correspondingly. Then, Section 5 explores the STBC-based enhancement amongst different modulation formats. In Section 6, we further investigate the bit error rate (BER) versus optical signal-to-noise ratio (OSNR) curves of mode-dependent loss equalization. Last but not least, Section 7 demonstrates the complexity optimization using the proposed single-stage architecture. Finally, Section 8 provides a summary of recent expressive experiments of SDM transmission using MIMO DSP for short-, medium- and long-haul distances.

2. Principle

This section introduces the theoretical background of MIMO equalizers, including LMS, RLS, and the space–time block-coding (STBC) algorithm for SDM optical transmission. Moreover, we also cover the aspects of polarization-related dynamic equalization, orthogonal frequency-division multiplexing (OFDM) FDE requirements, and the MIMO hardware requirements in this section. Although hypothetically any modal cross-talk can be electronically managed by DSP, the algorithmic complexity of the MDM coherent receiver remains one of the main limiting factors to the MIMO equalization scheme. In general, the total complexity of the DSP algorithm can be measured by the number of complex multiplications per symbol per mode, for the reason that the complex multipliers are usually the most resource-consuming arithmetic logics in terms of power consumption, area, and cost in application-specific integrated circuit (ASIC) design [21].

2.1. MIMO Equalizers: State of the Art

Although the SDM conception was present for quite a long time, the real SDM transmission systems were not pursued in industry until recent years [22]. This is because, in MDM systems, it is assumed that signals could be immaculately conveyed under a particular mode without interfering with other spatial channels within an ideal FMF; however, in reality, that might not be the case, for the signals in numerous modes are often cross-coupled to each other due to bending, twisting, and/or fiber fabrication imperfections; henceforward, the orthogonality of modes might only be preserved for a very short distance in industrial SDM applications.

In order to resolve such issue, MIMO signal processing was developed to undo channel cross-talk and recover the transmitted data quality. The coherent receiver structure for a single-carrier system with a 6 × 6 MIMO FDE is exemplified in Figure 3, where the coefficient adaptation can be accomplished with LMS or RLS algorithms along with data-aided training.

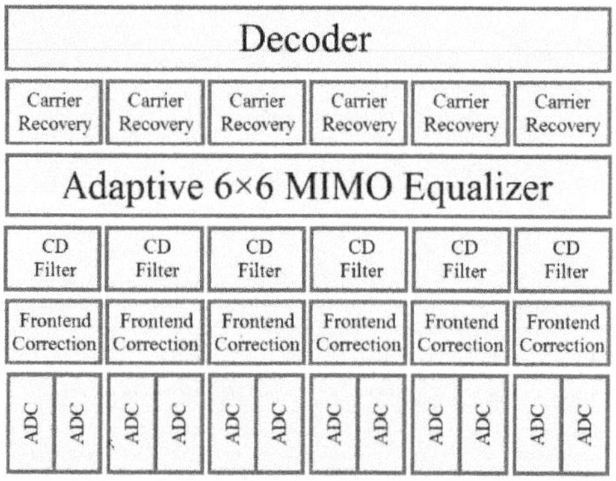

Figure 3. The typical coherent receiver configuration for a single-carrier transmission system with a 6 × 6 multi-input multi-output (MIMO) frequency-domain equalization (FDE) [7].

The conventional time-domain equalization (TDE) uses a finite impulse response (FIR) filter matrix adapted by the LMS algorithm. The complexity for the TDE C_{TDE} is given by [23]

$$C_{TDE} = 3 \cdot m \cdot \Delta\beta_1 \cdot L \cdot R_S, \quad (1)$$

where $\Delta\beta_1$ indicates the DMGD of the fiber, m signifies the number of modes within the fiber, L represents the length of the fiber, and R_S denotes the symbol rate. Subsequently, the complexity of TDE scales linearly with the mode group delay of the optical link, the number of modes, and the symbol rate. Meanwhile, the complexity for FDE C_{FDE} can be written in the form of

$$C_{FDE} = 4 \cdot (m+1) \cdot \log_2(2 \cdot \Delta\beta_1 \cdot L \cdot R_S) + 8 \cdot m. \quad (2)$$

It can be resolved from the equations above that FDE might substantially reduce the complexity compared with TDE. Characteristically, a block length equal to twice the equalizer length is applied for the adaptive FDE. By dint of even and odd sub-equalizers, an equivalent half-symbol period delay spacing finite impulse response (FIR) filter can be realized in the frequency domain. The required MIMO equalizer duration to compensate for DMGD in the weak-coupling regime can be transcribed as

$$N_{DMGD} = \Delta\beta_1 \cdot L \cdot R_{OS} \cdot R_S, \quad (3)$$

where ... means the ceiling function, $\Delta\beta_1$ signifies the DMGD in the fiber, and R_{OS} represents the oversampling rate. Furthermore, the MDL can be expressed as [24]

$$\text{MDL} = \max_{k=1\sim N}\left(\lambda_k^2\right) / \min_{k=1\sim N}\left(\lambda_k^2\right), \quad (4)$$

where λ_k stands for singular values of the coupler transfer matrix, which can be calculated by employing the singular value decomposition (SVD).

To achieve the MIMO equalization, the impulse-response matrix measurement needs to be performed first. An example of the squared magnitude of the polarization-division multiplexing (PDM) SDM 6 × 6 impulse responses for 96 km of a six-mode FMF is depicted in Figure 4, where the columns correspond to the transmitted ports and the rows correspond to the received ports [25]. In Figure 4, the sharp peaks in the subplots identify the main coupling points evidently, which are divided into four regions designated with A (the coupling between the polarization modes LP_{01x} and LP_{01y}), B (the coupling between the spatial and polarization modes LP_{11ax}, LP_{11ay}, LP_{11bx}, and LP_{11by}), and C and D (the cross-talk between LP_{01} and LP_{11} modes). Such channel estimation provides a clear picture of the cross-talk introduced by the mode multiplexer and the propagation through the six-mode FMF, which allows a better understanding of the MIMO DSP performance.

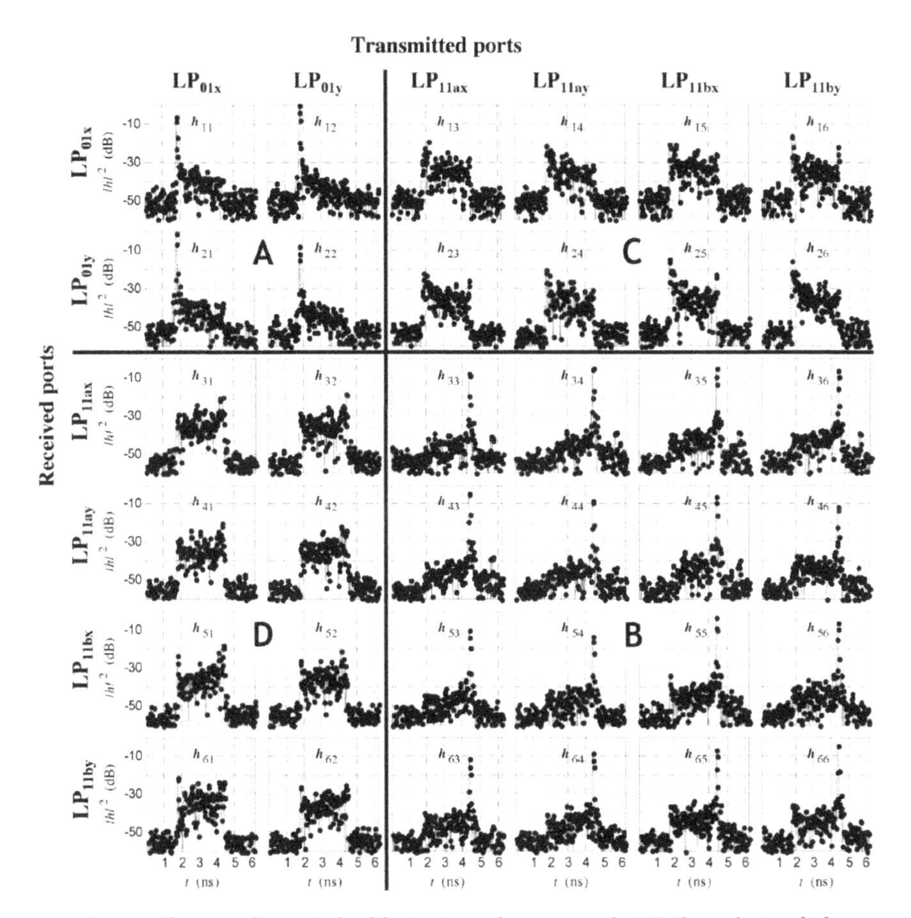

Figure 4. The squared magnitude of the 6 × 6 impulse responses for MIMO equalization [25].

Furthermore, a typical eye diagram is shown below of two de-correlated 12.5-Gbps NRZ 2^7-1 pseudorandom binary sequences (PRBSs) launched into one fundamental mode LP_{01} and a higher-order LP_{12} mode, with power adjusted such that both channels received the same power of −12 dBm at the receiver, with decent isolation, low los,s and minimal modal mixing [26]. Consequently, we can observe any substantial coupling between the modal groups within the FMF which would easily close the eye at the receiver. The eye diagram with heavy modal dispersion can be seen in Figure 5b, while the eyes of multiplexing independent channels LP_{01} and LP_{12} are displayed in Figure 5c,d.

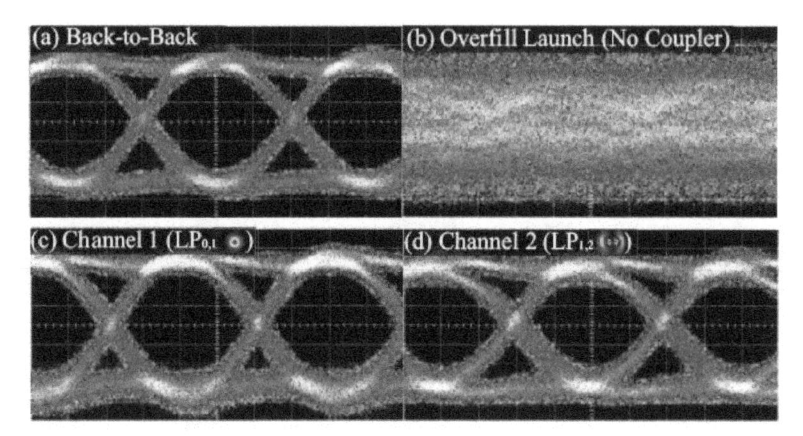

Figure 5. Eye diagrams of a typical mode-division multiplexing (MDM) transmission system for (**a**) back-to-back; (**b**) over-fill launch without coupler; (**c**) LP_{01} channel; (**d**) LP_{12} channel [26].

2.2. LMS and RLS Techniques

Both LMS and RLS algorithms are intended to iteratively minimalize the squared error at all discrete frequencies. The LMS algorithm stands as a stochastic gradient descent minimization using instantaneous error estimates, with the corresponding weight update as [27]

$$W(k+1) = W(k) + \mu[U(k) - W(k)V(k)], \tag{5}$$

where $W(k)$ signifies the MIMO channel matrix vector at frequency k, $U(k)$ indicates fast Fourier transform (FFT) of detected data blocks, μ denotes the step size, and $V(k)$ represents the FFT of sampled received signals.

In conventional LMS, the convergence rate and equalization performance depend on the same value of the scalar step size μ. The convergence speed can be enlarged by choosing an adaptive step size based on the power spectral density (PSD) of input signal, which can be conveyed as [28]

$$\mu(k) = \frac{\alpha}{S(k)}, \tag{6}$$

where S(k) signifies the PSD of a posterior error block, and α symbolizes the adaptation rate, which determines both the convergence speed and equalization performance of the noise PSD directed (NPD) LMS algorithm. The basic schematic of an MDM transmission system using the LMS algorithm is shown in Figure 6.

In contrast, the RLS algorithm depends on the iterative minimization of an exponentially weighted cost function, whose convergence speed is not strongly dependent on the input statistics, with the corresponding weight update as

$$W(k+1) = W(k) + [U(k) - W(k)V(k)] \cdot V(k)^H \cdot \left[R(k) \cdot \beta^{-1}\right], \tag{7}$$

where R(k) symbolizes a tracked inverse time-averaged weighted correlation matrix, the superscript H signifies the Hermitian conjugate, and β denotes a forgetting factor, which gives an exponentially lower weight to older error samples. Consequently, the main difference between the conventional LMS and RLS algorithms is that RLS has a growing memory due to R(k) and β, and, thus, might achieve an even lower OSNR and faster adaptation. Specifically, the tracked inverse time-averaged weighted correlation matrix can be denoted as [29]

$$R(k+1) = \left[R(k)\cdot\beta^{-1}\right] - \left\{\frac{\left[R(k)\cdot\beta^{-1}\right]\cdot V(k)\cdot V(k)^{H}\cdot\left[R(k)\cdot\beta^{-1}\right]}{1+V(k)^{H}\cdot\left[R(k)\cdot\beta^{-1}\right]\cdot V(k)}\right\}, \tag{8}$$

which assigns a different step size to each adjustable RLS equalizer's coefficient at their updates, rendering all frequency bins with a uniform convergence speed, thus cultivating the total algorithmic performance.

2.3. STBC-Assisted RLS Algorithms

One of the primary issues regarding the conventional RLS implementation might be its comparatively sophisticated computational complexity, attributable to the recursive updating with a growing memory. Henceforth, in order to accomplish a quasi-RLS performance at a quasi-LMS cost, the STBC technique can be applied to aid the RLS approach, which is defined as follows: assuming multiple copies of data stream are transmitted across a number of spatial modes in an FMF transmission, some of the received copies are better than the others under the same scattering, reflection, or thermal noise; thus, the various received versions of the data can be exploited to increase the overall system reliability [30].

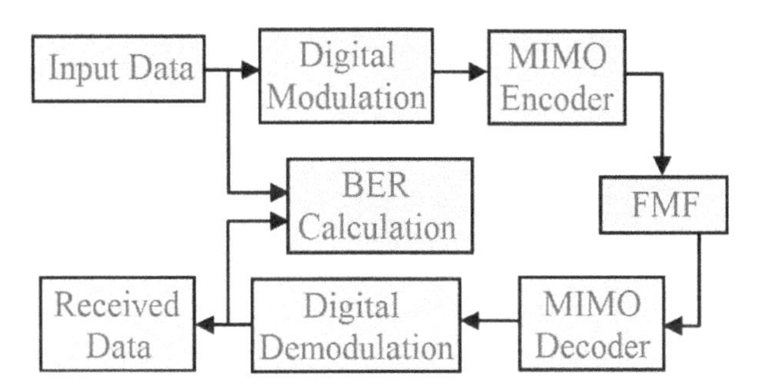

Figure 6. Schematic of an MDM system utilizing the MIMO FDE [31].

The space and time allocation for the STBC-aided RLS algorithm can be briefly represented as shown in Figure 7, as a matrix of spatial channels and time slots, whereby each element is the modulated symbol to be transmitted in mode *i* at time slot *j*. We may choose different combinations, and we can apply one or more of received copies to properly decode the received signal and, therefore, achieve better performance.

The architecture of the adaptive STBC-RLS equalizer in an m-mode FMF transmission system is illuminated in Figure 8. At coherent detection, the serial-to-parallel (S/P) converters split each data sequence $y_1, \ldots y_m$ into even/odd sequences, while two consecutive blocks are concatenated. After the FFT, the samples of the k-th block are overlapped with the (k − 1)th block at frequency k, which are further alienated into sequential blocks $y_q^P(k)$, where superscript P specifies an even or odd sequence. Each block is then converted to the frequency domain with FFT as $Y_q^P(k)$, which goes through an inversed channel filter for the data matrix production with the even or odd tributaries, while $H_{q,j}^P(k)$ stands for the inversed channel filter in Jones-vector notation, while both q and j denote mode indices between 1 and m. After an adaptive MIMO equalization, the carrier recovery is applied to mitigate the laser phase noise, the outputs of which are transformed back to the time domain using inverse FFT (IFFT), compared to the desired response to generate an error vector, and to update the receiver coefficients in a training mode until they converge. Then, it switches to a decision-directed mode, whereby reconstructed data are transformed back using FFT through the zero padding to update the coefficients [31].

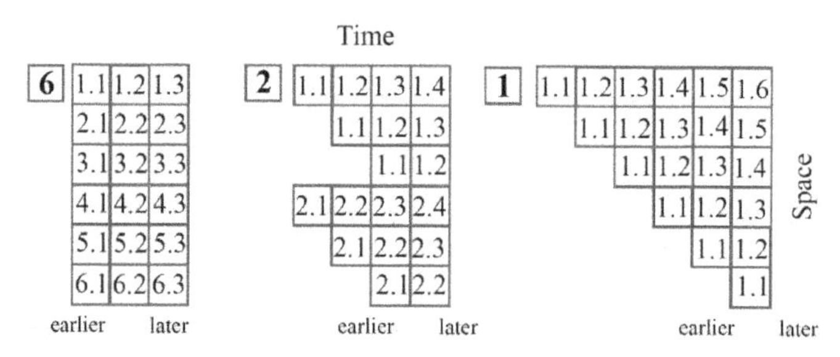

Figure 7. The space and time allocation for the space–time block-coding (STBC)-aided recursive least squares (RLS) algorithm [31].

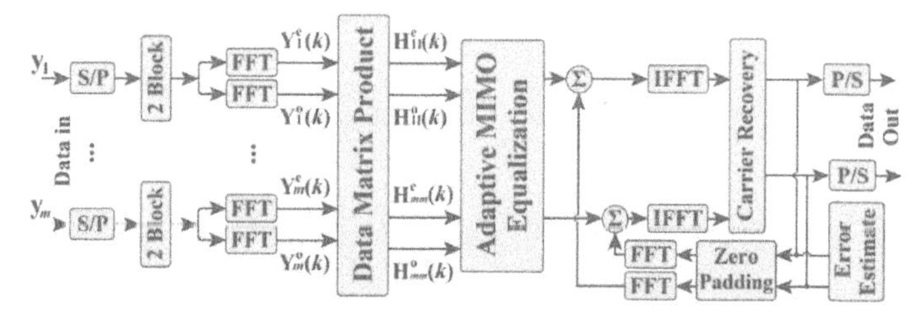

Figure 8. The architecture of an STBC-aided RLS equalizer for MDM receivers [29].

The weight update of the STBC-aided RLS algorithm follows the rules below, whereas the corresponding equalizer coefficients are updated every two blocks (k + 2) instead of each block (k + 1) in the recursion.

$$W(k+2) = W(k) + \begin{pmatrix} P_{k+2} & 0 \\ 0 & P_{k+2} \end{pmatrix} (U_{k+2})^{*} \cdot [D_{k+2} - U_{k+2} \cdot W(k)], \quad (9)$$

where U_{k+2} denotes the Alamouti matrix containing the received symbols from block k + 2 and k + 3, and D_{k+2} represents the desired response vector for training and decision-directed tracking [32], while the superscript * indicates the complex conjugation. Meanwhile, the diagonal term P_{k+2} is given by

$$P_{k+2} = \beta^{-1} \cdot \left[P_k - \beta^{-1} \cdot P_k \cdot \Omega_{k+2} \cdot P_k \right], \quad (10)$$

where β is the forgetting factor, and the term Ω_{k+2} is given by

$$\Omega_{k+2} = \left\{ \mathbf{Diagonal}\left[|V(k)|^2 + |V(k+1)|^2 \right]^{-1} + \beta^{-1} \cdot P_k \right\}^{-1}, \quad (11)$$

where V(k) stands for the FFT of sampled received signals. By utilizing the STBC scheme, we may choose different combinations, and apply one or more of received copies to correctly decode the received signal, and, thus, attain a better performance. Meanwhile, the STBC is able to accomplish an RLS-level performance with a much lower computational complexity, because no matrix inversion is essential during the recursion. Instead of adding a training sequence to each data block, only a few training blocks are added in the beginning, and the channel variations are tracked at the decision-directed stage in an attempt to reduce the overall system overhead [33].

2.4. Polarization-Related Dynamic Equalization

Polarization characterizes a fundamental multiplexing technique, which allows doubling the spectral efficiency of fiber optical transmission systems [34]. To accommodate the usage of two polarization tributaries, polarization de-multiplexing (PolDemux) algorithms represent a crucial part of digital coherent receivers, which makes it conceivable to efficiently compensate for time-varying state of polarization (SOP). The PolDemux approaches include MIMO algorithms such as LMS and RLS, and the constant-modulus algorithms (CMA).

In particular, Stokes space-based DSP techniques improve the polarization de-multiplexing, polarization dependent loss (PDL) compensation, and other polarization-related propagation impairments in coherent receivers, particularly in terms of convergence speed and transparency to higher-level M-ary modulated signals. The block diagram of the DSP subsystems of a coherent transceiver for adaptive equalization of both polarization and phase diversity is presented in Figure 9, which includes the front-end compensation, static equalization, adaptive equalization, carrier phase recovery, and the symbol decision [35].

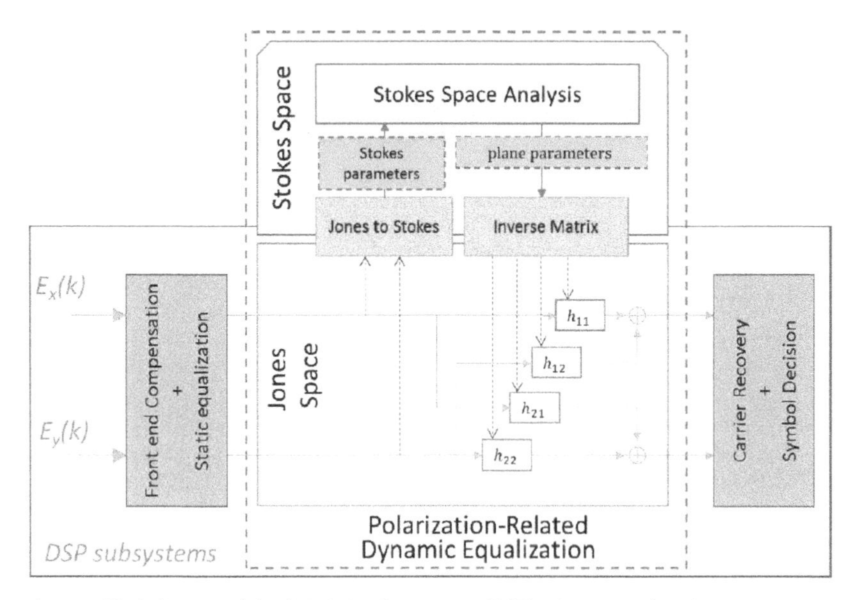

Figure 9. Block diagram of the digital signal processing (DSP) subsystems of a coherent transceiver using a Stokes space-based polarization de-multiplexing (PolDemux) algorithm [35].

Furthermore, the Stokes space-based PolDemux method has specific advantages including the higher convergence ratio, the improved robustness against phase noise, and the transparency to higher-level M-ary modulated signals.

2.5. OFDM FDE Requirements

The high-speed orthogonal frequency-division multiplexing (OFDM) in SDM optical long-haul transmission systems serves as another important aspect of MIMO equalization techniques [36]. The complex multiplications per bit for various equalizer types in terms of modal dispersion for a 2000-km transmission distance and 10×10 MIMO is shown in Figure 10, where an FMF with three modes with two polarizations (LP_{01}, LP_{11a}, LP_{11b}, LP_{21a}, LP_{21b}) results in the number of tributaries being 10.

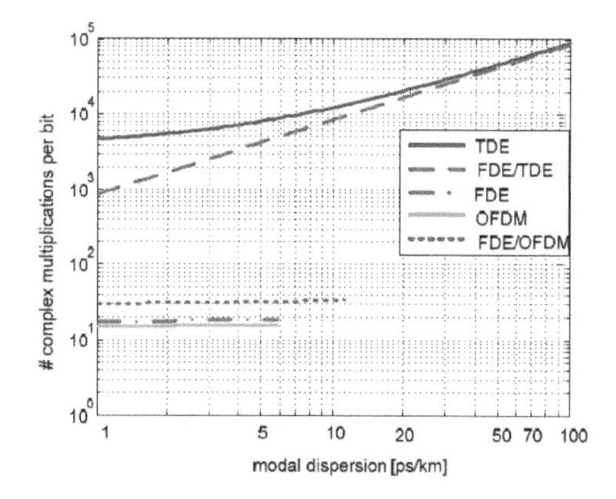

Figure 10. The orthogonal frequency-division multiplexing (OFDM) FDE equalizer complexity requirements compared with other MIMO approaches [37].

In principle, the complexity is calculated in a two-dimensional matrix, where one variable is the sampling rate and the other is the size of the FFT. For each configuration, the optimum sampling rate resulting in minimum complexity satisfying the overhead constraint is considered. As we can see, OFDM requires the lowest equalizer complexity for cross-talk compensation in an MDM receiver, for most of the multiplications required for FDE/TDE are caused by time domain operations. However, OFDM cannot tolerate a modal dispersion of more than 5.9 ps/km modal dispersion due to the 10% overhead constraint for FMF-based optical transmission [37].

2.6. MIMO Hardware Requirements

Last but not least, this subsection discusses the design and architecture of spatial multiplexing MIMO decoders for field-programmable gate arrays (FPGAs). Conventionally, the optimal hard-decision detection for MIMO systems is the maximum likelihood (ML) detector, which is well known due to its superior performance in terms of BER, but its direct implementation grows exponentially with a higher-level modulation scheme, causing its FPGA implementation to be relatively infeasible [38].

On the other hand, soft-output generation for a sphere detector can solve the ML detection issue in a computationally efficient manner, for a real-time implementation on a DSP processor at high-performance parallel computing platforms. The scalable architecture of soft-output generation for a sphere detector is presented in Figure 11, which requires further micro-architecture optimizations and trade-offs [39].

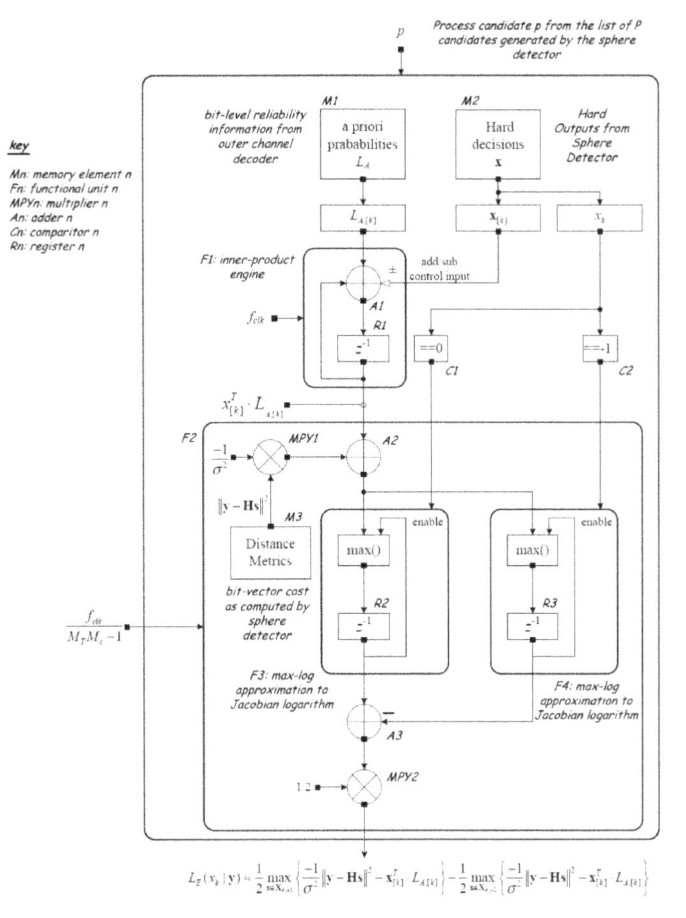

Figure 11. A scalable architecture of spatial multiplexing MIMO decoders for field-programmable gate arrays (FPGAs) [39].

3. System Configuration

The system configuration of a six-mode coherent transmission system is illustrated in Figure 12, with a $2^{23} - 1$ pseudorandom binary sequence (PRBS) sequence modulated on every mode over a 30-km FMF. The OSNR setting block indicates that the additive white Gaussian noise (AWGN) noise is added after the FMF transmission, while the optical filtering block specifies that a Gaussian filter with a 33-GHz bandwidth is used to suppress the out-of-band noise before the coherent detection.

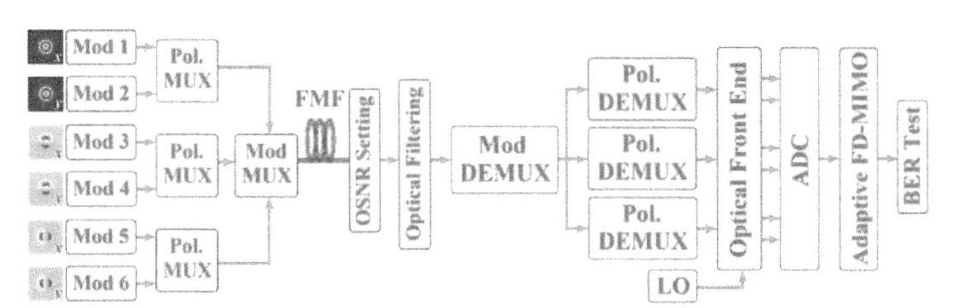

Figure 12. Configuration of a six-mode coherent transmission system [29].

After mode de-multiplexer (Mod DEMUX), parallel signals are launched into the receiver, and are converted to baseband in the optical front end by internally mixing with the local oscillator (LO) at a linewidth of 100 kHz. The electrical signal in-phase and quadrature (I/Q) components are then re-sampled at two samples per symbol by the analog-to-digital converter (ADC). After the MIMO processing, a BER is estimated for each mode [40].

4. Performance Evaluation for Different Adaptive Filtering Schemes

In this section, the convergence speed comparison amongst the conventional LMS algorithm, the noise PSD directed (NPD) LMS algorithm, the RLS algorithm, and the proposed STBC-RLS algorithm for PDM-QPSK signal is exemplified in Figure 13, whereas the normalized mean square error (NMSE) represents the average square errors of each block after the MIMO equalization. The adaptation rate and step size of four algorithms were set to their optimal values; the block length was 8192 samples using 50% overlap with OSNR set at 18 dB. In the meantime, the inset on the right displays the constellation diagram before the STBC compensation, as the received signal was severely distorted, attributable to mode coupling and delay, while the inset on the left indicates the constellation diagram after applying the STBC-RLS compensation, as the modal distortions were remarkably compensated for by the MIMO equalizers over a 30-km FMF.

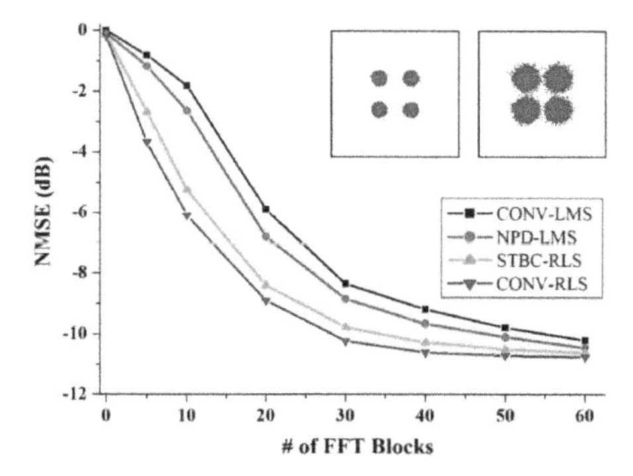

Figure 13. The normalized mean square error (NMSE) vs. the number of fast Fourier transform (FFT) blocks for various algorithms for a polarization-division multiplexing (PDM) quadrature phase shift keying (QPSK) signal; inset: constellation diagrams before (**right**) and after (**left**) the STBC-RLS compensation [27].

As for the computational complexity per data symbol for the MIMO FDE, which is mainly determined by the number of complex multiplications, the RLS and LMS algorithms would consume 148 and 302 complex multiplications, respectively, while the proposed STBC-RLS would require 172 complex multiplications, which is roughly a 16.2% increase in hardware complexity compare to LMS, but 75.6% less than that of the RLS approach.

To further upsurge the spectral efficiency of the next-generation SDM transmission systems, the convergence speed comparisons between different algorithms for PDM-16QAM and PDM-64QAM signals are presented in Figures 14 and 15, respectively, whereas the comparison between different modulation formats is summarized in Table 1.

From the plot above, we can see that RLS converges faster to a lower asymptotic NMSE than LMS, because it has a growing memory due to the forgetting factor. The NPD-LMS algorithm can achieve a faster convergence than the traditional LMS, because it adopts variable bin-wise step size to render posterior error of every frequency bin convergent to the background noise of the AWGN channels. Still, the NMSE convergence of the proposed STBC-RLS algorithm seems a bit inferior to that of the conventional RLS in the beginning, for a smaller block size is used for training than the data block size to reduce the system overhead at the expense of a minor loss of performance. Subsequently, to achieve −10 dB steady NMSE, which is equal to a 9.8-dB Q-value, the LMS and NPD-LMS schemes need 55 and 47 FFT blocks, while the RLS and STBC-RLS approaches require roughly 27 and 31 blocks, respectively; henceforward, the convergence rate would be enhanced by 50.9% and 43.6%, respectively.

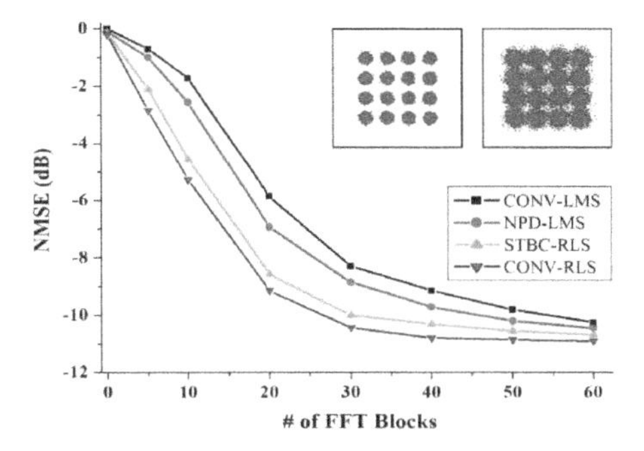

Figure 14. The NMSE vs. the number of FFT blocks for various algorithms for a PDM 16 quadrature amplitude modulation (16QAM) signal; inset: constellation diagrams before (**right**) and after (**left**) the STBC-RLS compensation [27].

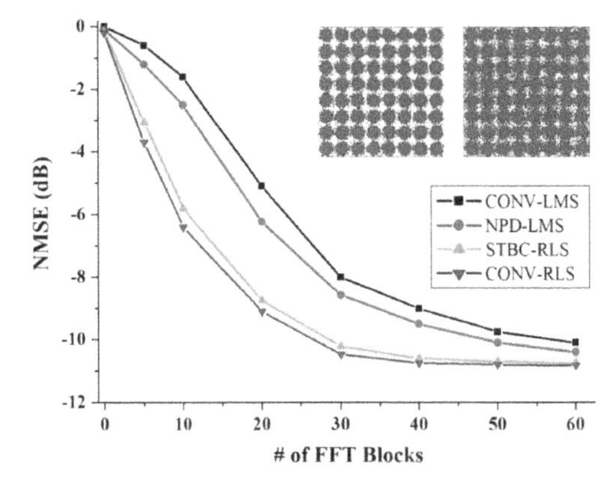

Figure 15. The NMSE vs. the number of FFT blocks for various algorithms for a PDM-64QAM signal; inset: constellation diagrams before (**right**) and after (**left**) the STBC-RLS compensation [27].

Table 1. Comparison of multiple modulation formats [21]. QPSK—quadrature phase shift keying; QAM—quadrature amplitude modulation.

Modulation Format	# of Symbols Transmitted	Levels per Carrier	# of Bits per Symbol Contained	Corresponding Data Rate
QPSK	$4 = 2^2$	2	2	112 Gb/s
16QAM	$16 = 2^4$	4	4	224 Gb/s
64QAM	$64 = 2^6$	8	6	336 Gb/s

From the plots, we can see that with higher-order modulation formats, the advantage of RLS convergence rate over that of the LMS becomes even larger, owing to a growing memory, while the difference in NMSE between the proposed STBC-RLS algorithm and the conventional RLS shrinks, which indicates that such proposed adaptive receivers could lower system overhead requirements for higher-order modulation formats.

5. STBC-Assisted Improvement between Various Modulation Formats

To irradiate the benefits of using the STBC scheme to mitigate the MDL impairment in the SDM transmission systems, the performance results of the space–time coded FMF transmission with or without using the STBC-RLS scheme are presented in Figure 16, from which we can achieve a roughly 3-dB OSNR improvement for the PDM-QPSK signal at a 10^{-3} BER, whereas the condition for the dotted line marked as Gaussian indicates the performance over a perfect MDL-free Gaussian channel.

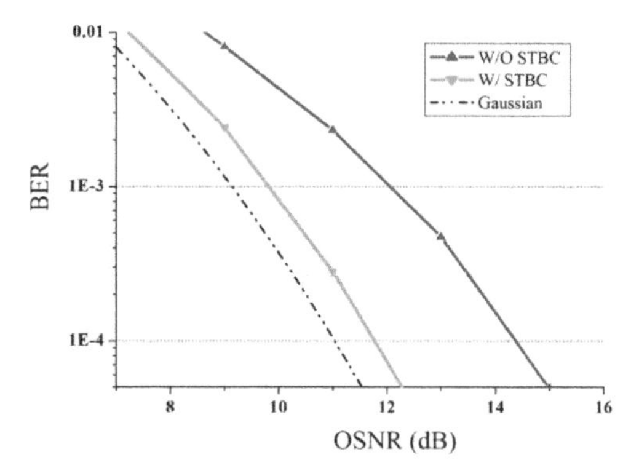

Figure 16. The bit error rate (BER) vs. optical signal-to-noise ratio (OSNR) comparison for PDM-QPSK signal in LP_{11a} mode with or without using the STBC-RLS scheme [31].

Additionally, for the higher-order modulation formats, the performance comparison plots for PDM-16QAM and PDM-64QAM signals with or without using the purposed STBC-RLS scheme are further shown in Figures 17 and 18, respectively.

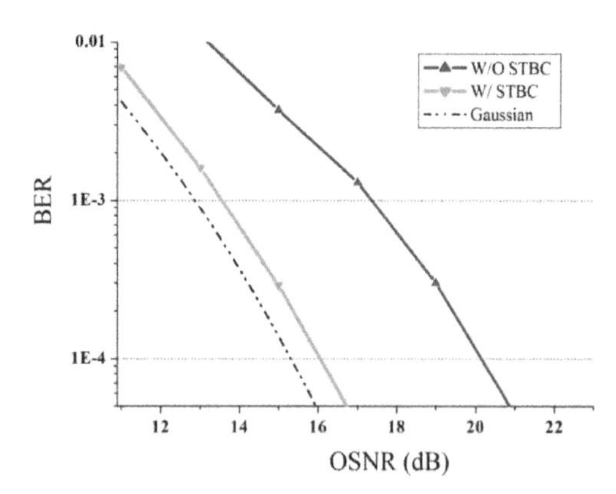

Figure 17. The BER vs. OSNR comparison for PDM-16QAM signal in LP_{11a} mode with or without using the STBC-RLS scheme [31].

In summary, the performance comparison for various modulation formats with or without using the STBC-RLS scheme is summarized in Figure 19, whereas the scheme for not using STBC-RLS designates the conventional RLS scheme, from which we can observe that, as more bits per symbol are transmitted, a larger OSNR tolerance improvement could be accomplished by using STBC for higher-order modulation formats. The overall OSNR tolerance can be improved by means of the STBC approach by approximately 3.1, 4.9, and 7.8 dB for QPSK, 16QAM, and 64QAM with respect to BER.

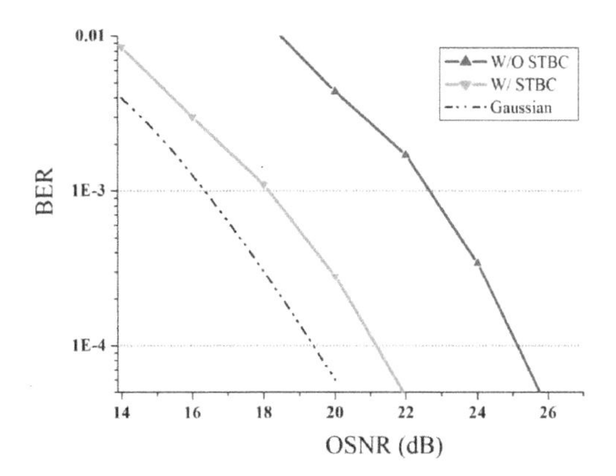

Figure 18. The BER vs. OSNR comparison for PDM-64QAM signal in LP_{11a} mode with or without using the STBC-RLS scheme [31].

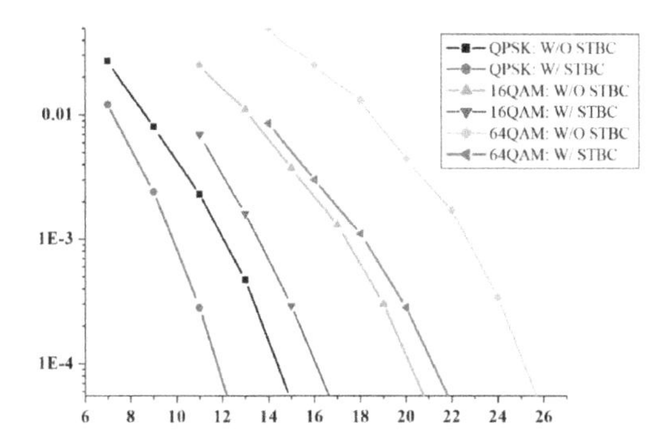

Figure 19. The summary of BER vs. OSNR for various modulation formats with or without using STBC-RLS compensation [29].

6. MIMO Equalization for Mode-Dependent Loss

The mode-dependent loss (MDL) arises from imperfections in optical components such as optical amplifiers, couplers, and multiplexers, as well as from non-unitary cross-talk in the fiber and at fiber splices and connectors, where the modes experience differential losses or gains when propagating through the optical link, which leads to signal-to-noise ratio disparities along with a loss of orthogonality between modes. The MDL is a capacity-limiting effect reducing the multiplexing benefit of SDM communication systems, which cannot be completely removed at the receiver by simply inserting mode scramblers. The probability distribution functions (PDF) of the accumulated MDL for the different coupling levels, with and without mode scrambling, are shown in Figure 20 as an example [41].

Furthermore, as there are up to three spatial modes being transmitted in our modeling, Figure 21 illustrates the equalization performance of the BER curves at different OSNRs by means of the STBC-RLS algorithm for a PDM-QPSK signal, from which we can see that the BERs of all spatial modes fall after MIMO equalization.

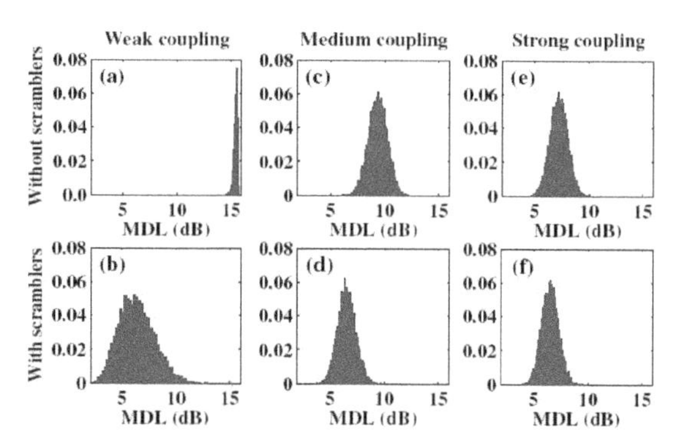

Figure 20. Mode-dependent loss (MDL) probability distribution functions under different coupling levels [41].

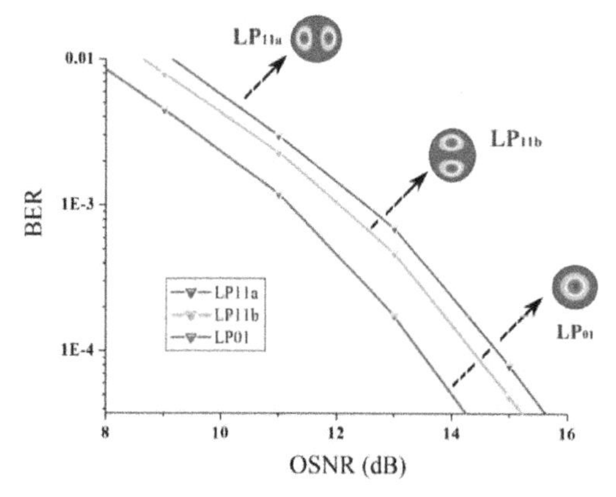

Figure 21. The BER vs. OSNR for three different modes for a PDM-QPSK signal using the STBC-RLS scheme [27].

The LP_{01} mode is able to realize a lower required OSNR for a 10^{-3} BER than the LP_{11} mode groups, predominantly because its more centralized mode pattern makes the LP_{01} mode suffer a lesser level of impact from mode coupling and cross-talk. Al though the LP_{11a} and LP_{11b} modes share an identical mode pattern as a pair of degenerate modes, their BER performance appears a bit dissimilar triggered by spatial mismatching and rotation. Since we found that the transmitted signals on both x- and y-polarizations provide a comparable performance, here, in these plots, the LP_{11b} mode is adopted as the average of two orthogonal polarizations for an easier analysis.

Additionally, the BER versus OSNR relationships based on the STBC-RLS algorithm for PDM-16QAM and PDM-64QAM signal compensation for LP_{01}, LP_{11a}, and LP_{11b} modes are presented in Figures 22 and 23, respectively. To reach a BER below 10^{-3}, the LP_{01} mode requires a 16.5-dB OSNR, while the LP_{11b} and LP_{11a} modes need about 17.7-dB and 18.1-dB OSNRs for the PDM-16QAM signal, while, for the PDM-64QAM signal, the LP_{01} mode requires a 21.3-dB OSNR, while the LP_{11b} and LP_{11a} modes need about 22.8-dB and 23.6-dB OSNRs, respectively.

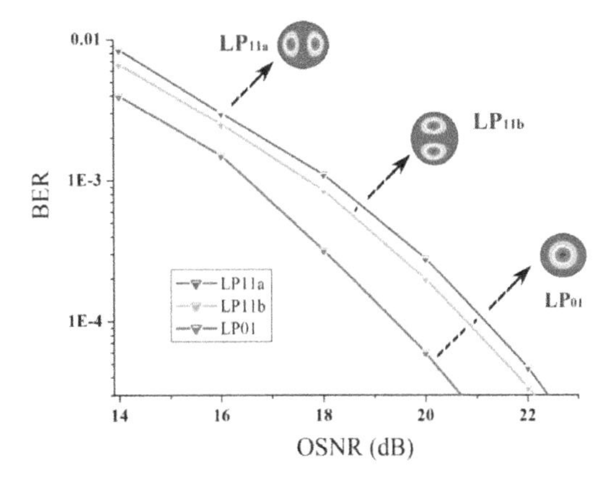

Figure 22. The BER vs. OSNR for three different modes for a PDM-16QAM signal using the STBC-RLS scheme [27].

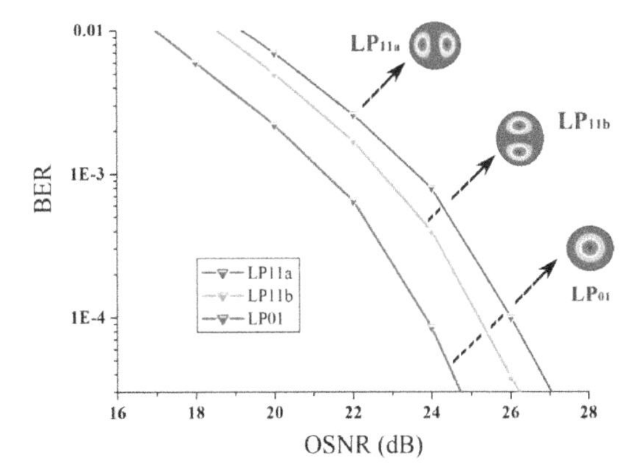

Figure 23. The BER vs. OSNR for three different modes for a PDM-64QAM signal using the STBC-RLS scheme [27].

7. Hardware Complexity Optimization for MIMO

This section focuses on the hardware complexity for MIMO equalization, which includes the FIR filter tap number for MIMO, single-stage architecture, and joint DSP for an MCF-based SDM. For a larger mode number m, frequency-domain (FD)-RLS might provide a better performance pertaining to the conventional FD-LMS algorithm, for the convergence speed of FD-RLS is not strongly dependent on the input statistics. Until now, the main issue of the adaptive FD-RLS implementation is its relatively higher computational complexity, due to the recursive updating with a growing memory. In this section, a joint chromatic dispersion (CD) and DMGD optimization technique is proposed for a further reduction in the algorithmic complexity of the FD-RLS approach.

7.1. Filter Tap Number for MIMO

The structure of an MIMO equalizer as an extension of the standard CMA is depicted in Figure 24, which is based on multiple FIR filters in a butterfly structure. The hardware complexity of such an architecture per transmitted bit is only doubled for the 4×4 MIMO configuration compared to standard single-mode operation, for sixteen adaptive filters are prerequisite to process 2×100 Gb/s with a 4×4 MIMO compared to four filters for 1×100 Gb/s with a 2×2 MIMO as a reference. On the right graph of Q^2-factor versus the FIR filter tap number, which is used to estimate the required length of the MIMO equalizer allowing the detection of all the polarization and mode tributaries, an almost flat factor from nine taps to 15 taps can be observed for both 100 Gb/s PDM-QPSK and 2 × 100 Gb/s MDM after a 40-km transmission [42].

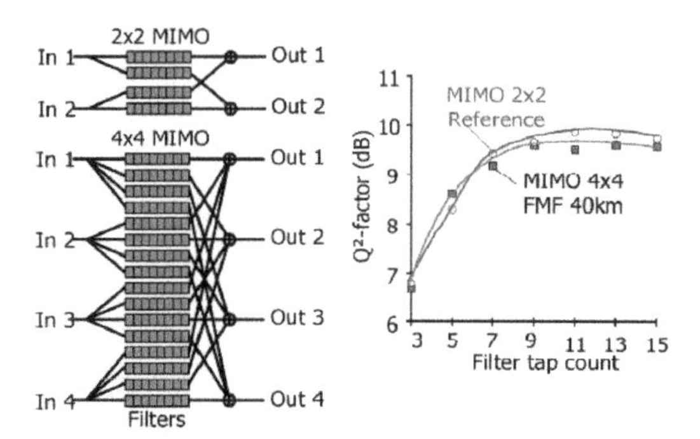

Figure 24. The finite impulse response (FIR) filter illustration, along with Q^2-factor vs. tap number for both 2 × 2 and 4 × 4 MIMO cases [42].

The required number of taps per carrier used as a function of distance in recent MDM transmission experiments is illuminated in Figure 25, where the DMGD increases linearly with distance in the weak coupling regime [19]. Fiber management is expedient in decreasing the maximum DMGD in an FMF. In addition, as different forms of signals, fibers, and spatial channels are likely to cohabit in future SDM transport networks, it is also significant that MIMO DSP has sufficient processing capability to handle all received signals.

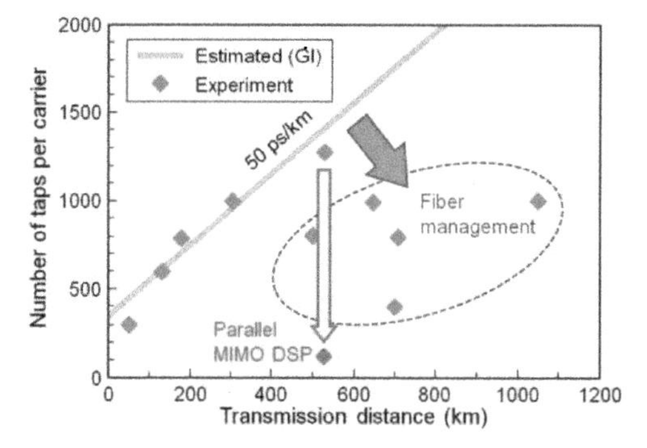

Figure 25. The number of taps per carrier for MIMO DSP vs. distance in recent few-mode fiber (FMF) transmission experiments [19].

7.2. Single-Stage Architecture

In this subsection, we discuss the complexity optimization using the single-stage architecture of CD and DMGD joint compensation. The equalization of CD and DMGD can be performed together using FFT for efficient implementation of convolution, as a favorable structure for digital coherent receivers, for sparing the CD compensation modules and, hence, reducing the total DSP implementation complexity [43].

The schematic diagram of a coherent receiver for the compensation of the linear channel effects such as CD and DMGD is exemplified in Figure 26. To attain a lower computational complexity, the equalization of CD and DMGD can be implemented together, compared with the conventional two-stage FDE structure, using FFT for an efficient implementation of convolution; thus, m independent static CD compensation modules would be spared. Subsequently, a joint MIMO FDE of the static CD along with the dynamic DMGD would be an auspicious configuration for digital coherent receivers, because it would save a separate CD compensation module, thereby decreasing the total DSP implementation complexity.

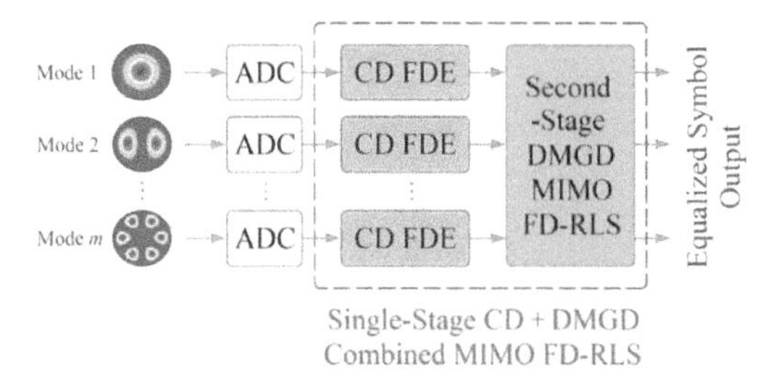

Figure 26. The representation of a coherent receiver with the joint compensation of chromatic dispersion (CD) plus differential mode group delay (DMGD) through a single-stage MIMO architecture [10].

Above and beyond, Figure 27 provides the comparison of the required complex multiplications per output symbol at different transmission distances for the two-stage and single-stage FD-RLS algorithms. For a transmission distance rising from 100 km up to 1000 km, the single-stage method continuously consumes fewer complex multiplications than the conventional two-stage method, because both methods use the same FFT size for DMGD compensation for the same transmission distance, whereas the FFT size of static FDE in the two-stage method is determined by the FMF-induced CD. For instance, the two-stage FD-RLS method consumes 323 and 351 complex multiplications per output symbol at 400-km and 800-km transmissions, whereas our proposed single-stage method takes only 297 and 328 complex multiplications per output symbol at the same transmission distances, respectively.

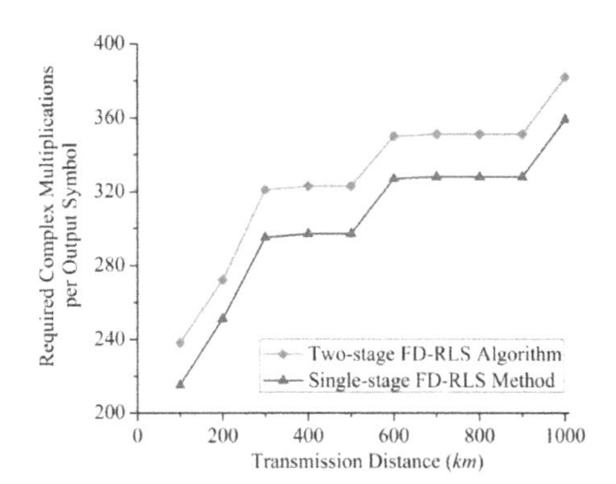

Figure 27. The required complex multiplications per output symbol of the two-stage and single-stage frequency-domain (FD)-RLS approaches as a function of transmission distance in an FMF-based transmission system [10].

Furthermore, Figure 28 irradiates the Q-value versus the step size of adaptive FD-RLS algorithms in both two-stage and single-stage methods for compensating for DMGD and CD. To attain a <0.5-dB Q-penalty at 400-, 700-, and 1000-km transmissions, the maximum step sizes are 4.5×10^{-5}, 3×10^{-5}, and 1.5×10^{-5}, respectively. In addition, for the same transmission distance, the maximum step sizes required to achieve the same Q-value after equalization are identical for both approaches.

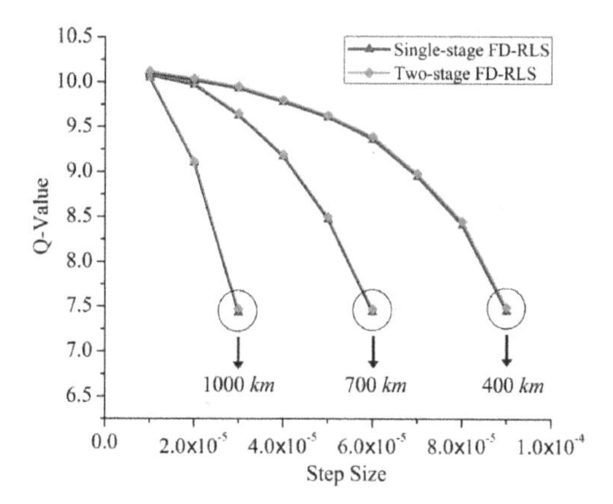

Figure 28. The Q-value vs. step size using both two-stage and single-stage FD-RLS algorithms as a function of transmission distance in an FMF-based transmission system [15].

Last but not least, Figure 29 exhibits the convergence speed comparison among three algorithms at different OSNR levels, from which we learnt the time required for convergence grows more slowly for FD-RLS than FD-LMS algorithms as the OSNR increases, equivalent to plummeting the overall training sequence overhead by approximately 30%.

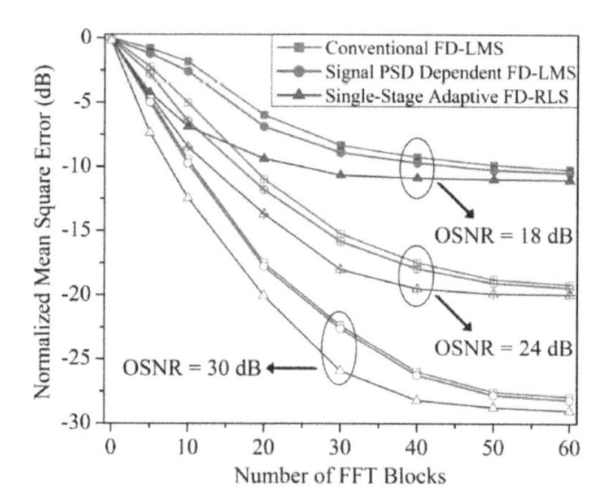

Figure 29. The convergence speed comparison of two FD-LMS algorithms and single-stage FD-RLS algorithms at different OSNRs [7].

7.3. Joint DSP for MCF-Based SDM

Last but not least, this subsection focuses on a joint DSP scheme for the MCF-based SDM transmission system to reduce the overall cost and power consumption of integrated receivers. There is a strong correlation between the phase fluctuations of the different sub-channels, and the master–slave phase recovery causes no BER degradation. The block diagram of a joint DSP method for an MCF-based SDM transmission system with master–slave phase recovery is presented in Figure 30, which utilizes the phase recovered from a single "master" sub-channel to eradicate phase recovery blocks in the "slave" sub-channels, thus decreasing the overall DSP burden at the receiver side [44].

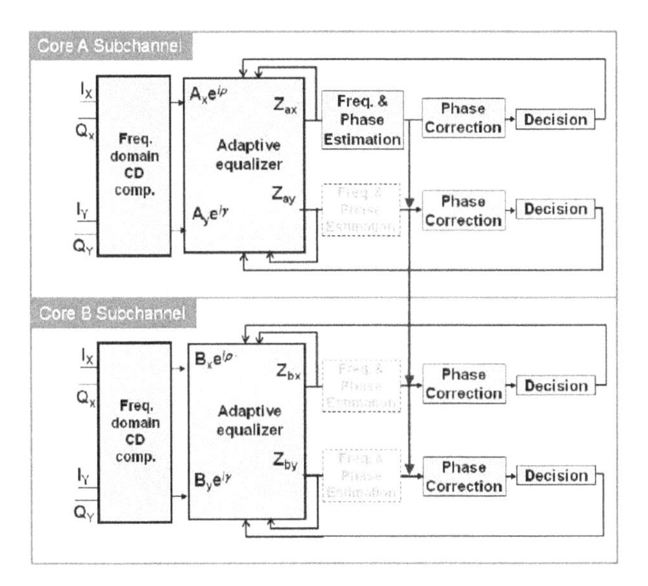

Figure 30. A joint DSP approach for multicore fiber (MCF)-based SDM transmission system with master–slave phase recovery [44].

In addition, the BER versus OSNR observed for master–slave phase recovery using the joint DSP scheme for MCF-based SDM transmission system is presented in Figure 31, which is equivalent to that found with independent phase recovery [45].

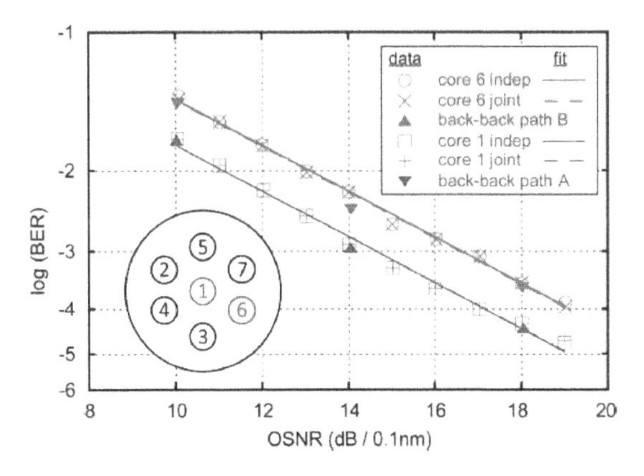

Figure 31. BER vs. OSNR for MCF-based SDM transmission system using the joint DSP scheme [45].

8. Recent Experiments of SDM Transmission Using FMFs

Last but not least, a couple of topical experimental verifications are assessed facilitating a higher transmission capacity for short, medium, and long distances in this section, utilizing FMFs and/or MCFs, along with the progressive maturity of the SDM amplifier, spatial mode coupler, and digital MIMO equalizers to safeguard higher capacity and spectral efficiency.

For instance, Ryf et al. conducted an experiment in 2018 for a MIMO-based 36-mode (two polarization modes) transmission over a 2-km-long 50-µm graded-index multimode fiber (MMF) with a spectral efficiency of 72 bit/s/Hz, as denoted in Figure 32 along with the close-ups of the heterodyne receiver arrangement and multimode acousto-optic modulator, as well as the spectrum of the test signal and location of the heterodyne filters and local oscillator [46]. The results were attained using a 72 × 72 coherent MIMO-DSP, with five 15-Gbaud dual-carrier PDM-QPSK signals with a channel spacing of 50 GHz transmitted.

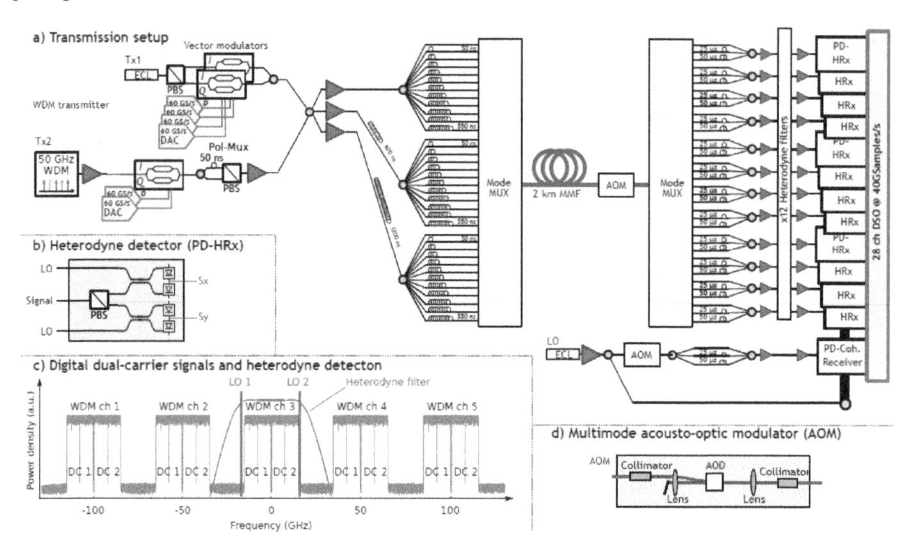

Figure 32. (**a**) Setup of a 72 × 72 MIMO-based transmission; (**b**) heterodyne receiver arrangement; (**c**) spectrum of the test signal and location of the heterodyne filters and local oscillator; (**d**) multimode acousto-optic modulator [46].

Figure 33 denotes the experimental set-up of an MIMO-based 45-mode (two polarization modes) MDM transmission over a 26.5-km-long 50-µm graded-index MMF over 20 wavelength channels resulting in a total capacity of 101 Tb/s and a spectral efficiency of 202/bit/s/Hz. The received 90-degree complex amplitude signals were processed by a 90 × 90 MIMO-FDE with 300 symbol-spaced taps, whereas the initial convergence of the equalizer was obtained using the data-aided FD-LMS algorithm with CMA used after that. Then, the carrier-phase recovery and BER counting were performed, while the Q-factors were computed by evaluating an inverse Q function of the BER [47]. The corresponding plots also include the Q-factors of 90 spatial tributaries for QPSK and 16QAM transmission in line with the mode groups, sorted by the performance within the mode group in Figure 33b, with the intensity-averaged impulse response attained from channel estimation over the MMF in Figure 33c, and the intensity transfer matrix showing the coupling between the mode groups in Figure 33d, along with the average Q-factor for QPSK and 16QAM as a function of wavelength in Figure 33e. The error bars within the plots signify the best and the worst spatial tributaries.

In another example, a 138 Tbit/s coherent transmission over six spatial modes plus two polarizations, and 120 wavelength channels carrying 120 Gbit/s mapped 16QAM signals over a 65 km FMF12 fiber are shown in Figure 34, where low insertion and MDL mode-selective photonic lanterns are used along with a coherent 12 × 12 MIMO equalization using a data-aided FD-LMS algorithm for resolving the modal mixing [48]. The FMF12 fiber is a special FMF designed to have low attenuation for the first 12 spatial channels, while having sufficiently high losses for the higher-order modes to guarantee an effective cut-off. At the receiver side, optical front-end impairment, CD, and frequency-offset compensation is performed, while the carrier phase recovery is performed before calculating performance metrics over 8 million bits per spatial channel. Moreover, the Q-factors versus the launch power averaged for 15 WDM channels, as well as for 120 WDM channels after 650-km transmission are shown in Figure 35, where the gradient of the shaded areas epitomizes the variation in WDM channel performance [48]. In addition, the larger performance variations on the edges of the spectrum, and the singular values for each WDM channel averaged, corresponding to the measured MDL, are depicted in Figure 35c,d, respectively.

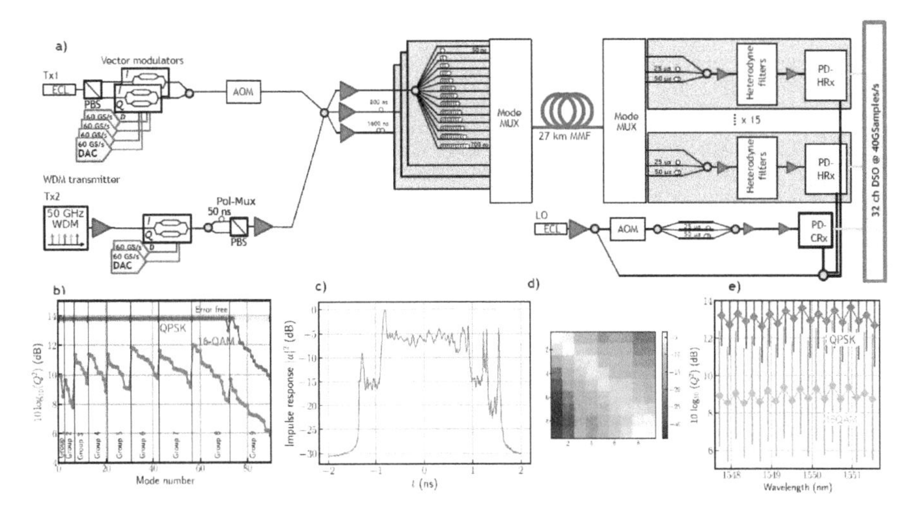

Figure 33. (**a**) Setup of a 90 × 90 MIMO-based transmission; (**b**) Q-factors for QPSK and 16QAM; (**c**) intensity-averaged impulse response attained; (**d**) intensity transfer matrix; (**e**) average Q-factor for QPSK and 16QAM [47].

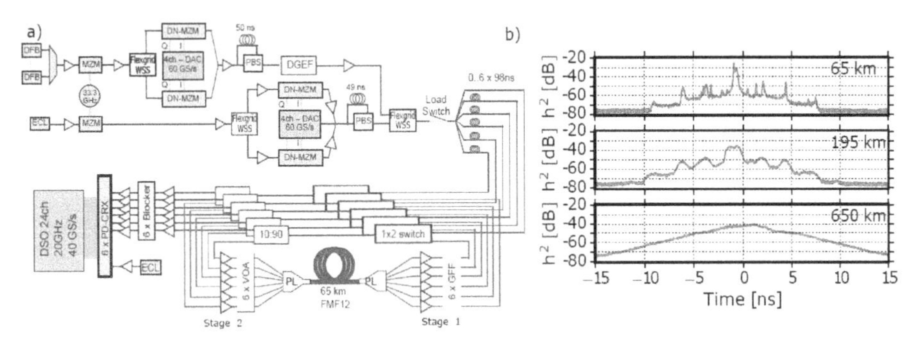

Figure 34. (**a**) Set-up of a 12 × 12 MIMO-based transmission; (**b**) impulse response after 65-, 195-, and 650-km transmission [48].

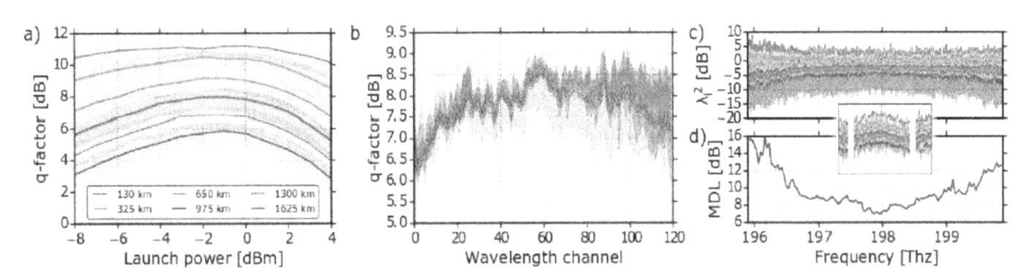

Figure 35. (**a**) Q-factors vs. power for 15 WDM channels; (**b**) Q-factors vs. WDM channels; (**c**) singular values after 650-km transmission; (**d**) average MDL vs. frequency per WDM channel [48].

Last but not least, the set-up of another 6 × 6 MIMO-based SDM/WDM transmission system using a 70-km DGD-compensated three-mode fiber is shown in Figure 36, whereas the received signals are off-line processed by firstly re-sampling the signals to two samples per signal, along with CD and frequency-offset compensation, which is then followed by a 6 × 6 MIMO FDE with 600 symbol-spaced taps using a data-aided LMS algorithm along with carrier-phase recovery and BER counting for Q-factor calculations [49]. The Q-factors of the center channel at 1550 nm for 64QAM, 16QAM, and QPSK signals as a function of distance for full space-division multiplexed, quasi-single-mode, and single-mode fiber transmission, respectively, are presented in Figure 37, with the markers indicating the measured distances [50]. A summary of these recent experimental validations using MIMO DSP is presented in Table 2 while an updated graph of mode numbers versus the transmission distance is presented in Figure 38, which summarizes the state-of-the-art SDM-WDM experiments [51–53]. Further improvement in the near future will most likely come from the soft-output maximum-likelihood algorithms, including repeated tree search and single tree search, to further improve the efficiency of the MIMO DSP [54].

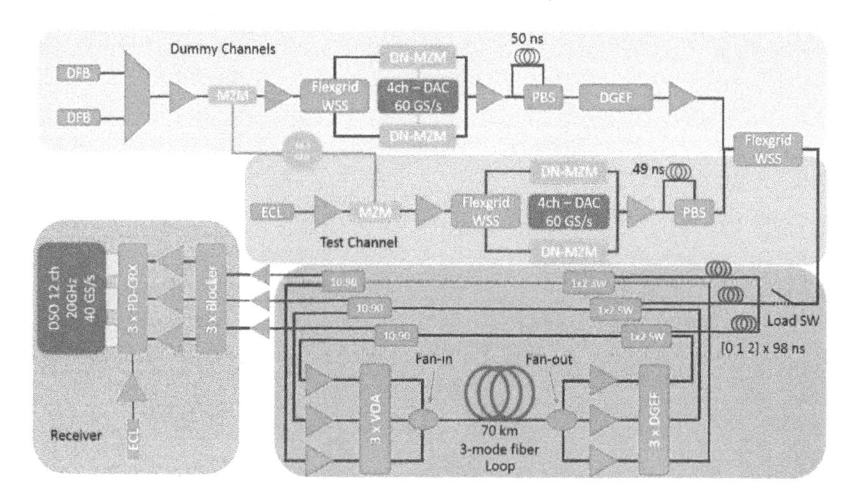

Figure 36. Setup of a 6 × 6 MIMO-based transmission [49].

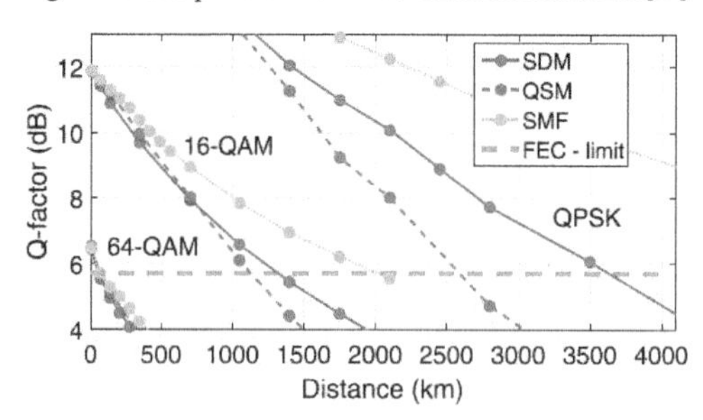

Figure 37. Q-factors for 64QAM, 16QAM, and QPSK [50].

Table 2. Summary table of recent experimental validations using multi-input multi-output (MIMO) digital signal processing (DSP). MMF—multimode fiber; FMF—few-mode fiber; MCF—multicore fiber.

Transmission Distance	Spectral Efficiency	DSP Approach	Fiber Type	Number of Modes	Year	Reference
2 km	72 bit/s/Hz	72 × 72 MIMO	50 μm MMF	36 × 2	2018	[46]
26.5 km	202 bit/s/Hz	90 × 90 MIMO	50 μm MMF	45 × 2	2018	[47]
65 km	34.91 bit/s/Hz	12 × 12 MIMO	FMF12	6 × 2	2017	[48]
70 km	9 bit/s/Hz	6 × 6 MIMO	FMF	3 × 2	2017	[49]
9.8 km	947 bit/s/Hz	6 × 6 MIMO	FMF + MCF	3 × 2	2016	[51]
23.8 km	43.63 bit/s/Hz	30 × 30 MIMO	FMF9	15 × 2	2015	[52]
53.7 km	217.6 bit/s/Hz	6 × 6 MIMO	MCF	3 × 2	2016	[53]

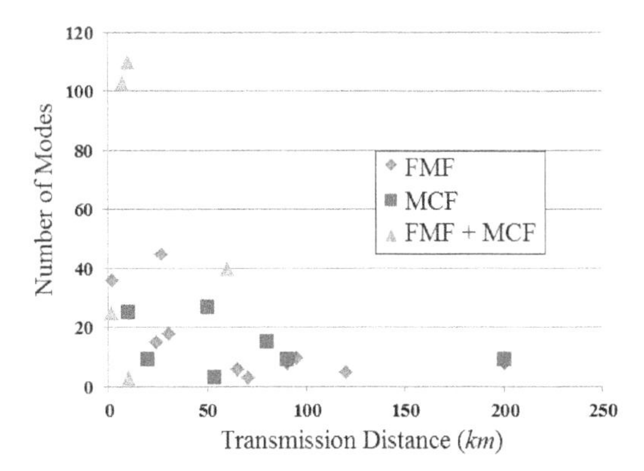

Figure 38. Updated graph of mode numbers vs. the transmission distance in the state-of-the-art SDM-WDM experiments.

9. Conclusions

In this review paper, we presented an overview regarding the state-of-the-art adaptive MIMO equalization for high-speed MDM coherent receivers, including the LMS and RLS techniques, as well as the STBC-assisted RLS algorithms. The performance analysis of various adaptive filtering schemes was presented for QPSK, 16QAM, and 64QAM, along with the STBC-based enhancement amongst different modulation formats. Last but not least, the complexity optimization using the single-stage architecture of CD and DMGD joint compensation was discussed, for reducing the total DSP implementation complexity. Additionally, a number of recent experimental validations were assessed, enabling higher transmission capacity and spectral efficiency for different transmission distances by means of various MIMO DSP approaches.

Author Contributions: Y.W. leads the development of algorithms and results used in the paper. J.W. contributes to the development of simulation mode and results. Z.P. supervises overall project.

Funding: This research received no external funding

Acknowledgments: We would like to thank the unknown reviewers that helped improve this manuscript with their valuable feedback.

Conflicts of Interest: The authors declare no conflicts of interest.

References

1. Winzer, P.J. Making spatial multiplexing a reality. *Nat. Photonics* **2014**, *8*, 345–348. [CrossRef]
2. Li, A.; Chen, X.; Amin, A.A.; Ye, J.; Shieh, W. Space-division multiplexed high-speed superchannel transmission over few-mode fiber. *J. Lightw. Technol.* **2012**, *30*, 3953–3964. [CrossRef]
3. Richardson, D.J.; Fini, J.M.; Nelson, L.E. Space-division multiplexing in optical fibres. *Nat. Photonics* **2013**, *7*, 354–362. [CrossRef]

4. Ip, E. Recent advances in mode-division multiplexed transmission using few-mode fibers. In Proceedings of the Twenty-Third National Conference on Communications (NCC), Chennai, India, 2–4 March 2017.
5. Zhang, Z.; Guo, C.; Cui, L.; Mo, Q.; Zhao, N.; Du, C.; Li, X.; Li, G. 21 spatial mode erbium-doped fiber amplifier for mode division multiplexing transmission. *Opt. Lett.* **2018**, *43*, 1550–1553. [CrossRef] [PubMed]
6. Jain, S.; Rancaño, V.J.F.; May-Smith, T.C.; Petropoulos, P.; Sahu, J.K.; Richardson, D.J. Multi-element fiber technology for space-division multiplexing applications. *Opt. Express* **2014**, *22*, 3787–3796. [CrossRef] [PubMed]
7. Pan, Z.; Weng, Y.; He, X.; Wang, J. Adaptive frequency-domain equalization and MIMO signal processing in mode division multiplexing systems using few-mode fibers. In Proceedings of the Signal Processing in Photonic Communications 2016 (SPPCom)—SpW2G.1, Vancouver, BC, Canada, 18–20 July 2016.
8. Arık, S.Ö.; Askarov, D.; Kahn, J.M. Adaptive frequency-domain equalization in mode-division multiplexing systems. *Opt. Express* **2014**, *32*, 1841–1852. [CrossRef]
9. Faruk, M.S.; Kikuchi, K. Adaptive frequency-domain equalization in digital coherent optical receivers. *Opt. Express* **2011**, *19*, 12789–12798. [CrossRef] [PubMed]
10. Weng, Y.; He, X.; Pan, Z. Performance analysis of low-complexity adaptive frequency-domain equalization and MIMO signal processing for compensation of differential mode group delay in mode-division multiplexing communication systems using few-mode fibers. In Proceedings of the SPIE 9774, Next-Generation Optical Communication: Components, Sub-Systems, and Systems V, 97740B, San Francisco, CA, USA, 13 February 2016.
11. Zhu, C.; Tran, A.V.; Do, C.C.; Chen, S.; Anderson, T.; Skafidas, E. Digital signal processing for training-aided coherent optical single-carrier frequency-domain equalization systems. *J. Lightw. Technol.* **2014**, *32*, 4110–4120.
12. Pittalà, F.; Slim, I.; Mezghani, A.; Nossek, J.A. Training-aided frequency-domain channel estimation and equalization for single-carrier coherent optical transmission systems. *J. Lightw. Technol.* **2014**, *32*, 4247–4261. [CrossRef]
13. He, X.; Weng, Y.; Wang, J.; Pan, Z. Noise power directed adaptive frequency domain least mean square algorithm with fast convergence for DMGD compensation in few-mode fiber transmission systems. In Proceedings of the Optical Fiber Communication Conference (OFC)—Th3E.1, San Francisco, CA, USA, 9–13 March 2014.
14. Puttnam, B.J.; Sakaguchi, J.; Mendinueta, J.M.D.; Klaus, W.; Awaji, Y.; Wada, N.; Kanno, A.; Kawanishi, T. Investigating self-homodyne coherent detection in a 19 channel space-division-multiplexed transmission link. *Opt. Express* **2013**, *21*, 1561–1566. [CrossRef]
15. Weng, Y.; Wang, T.; Pan, Z. Fast-convergent adaptive frequency-domain recursive least-squares algorithm with reduced complexity for MDM transmission systems using optical few-mode fibers. In Proceedings of the Conference on Lasers and Electro-Optics (CLEO)—SW4F.6, San Jose, CA, USA, 5–10 June 2016.
16. Ryf, R.; Weerdenburg, J.V.; Alvarez-Aguirre, R.A.; Fontaine, N.K.; Essiambre, R.-J.; Chen, H.; Alvarado-Zacarias, J.C.; Amezcua-Correa, R.; Koonen, T.; Okonkwo, C. White Gaussian noise based capacity estimate and characterization of fiber-optic links. In Proceedings of the Optical Fiber Communication Conference (OFC)—W1G.2, San Diego, CA, USA, 11–15 March 2018.
17. Choutagunta, K.; Arik, S.Ö.; Ho, K.-P.; Kahn, J.M. Characterizing mode-dependent loss and gain in multimode components. *J. Lightw. Technol.* **2018**, *36*, 3815–3823. [CrossRef]
18. Andrusier, A.; Meron, E.; Feder, M.; Shtaif, M. Optical implementation of a space-time-trellis code for enhancing the tolerance of systems to polarization-dependent loss. *Opt. Lett.* **2013**, *38*, 118–120. [CrossRef]
19. Mizuno, T.; Takara, H.; Shibahara, K.; Sano, A.; Miyamoto, Y. Dense space division multiplexed transmission over multicore and multimode fiber for long-haul transport systems. *J. Lightw. Technol.* **2016**, *34*, 1484–1493. [CrossRef]
20. Ip, E.; Milione, G.; Li, M.-J.; Cvijetic, N.; Kanonakis, K.; Stone, J.; Peng, G.; Prieto, X.; Montero, C.; Moreno, V.; Liñares, J. SDM transmission of real-time 10GbE traffic using commercial SFP + transceivers over 0.5 km elliptical-core few-mode fiber. *Opt. Express* **2015**, *23*, 17120–17126. [CrossRef] [PubMed]
21. Pan, Z.; He, X.; Weng, Y. Hardware efficient frequency domain equalization in few-mode fiber coherent transmission systems. In Proceedings of the SPIE 9009, Next-Generation Optical Communication: Components, Sub-Systems, and Systems III, San Francisco, CA, USA, 1 February 2014.
22. Ferreira, F.; Jansen, S.; Monteiro, P.; Silva, H. Nonlinear semi-analytical model for simulation of few-mode fiber transmission. *IEEE Photonics Technol. Lett.* **2012**, *24*, 240–242. [CrossRef]

23. Poletti, F.; Horak, P. Description of ultrashort pulse propagation in multimode optical fibers. *J. Opt. Soc. Am. B* **2008**, *25*, 1645–1654. [CrossRef]
24. Mumtaz, S.; Essiambre, R.J.; Agrawal, G.P. Nonlinear propagation in multimode and multicore fibers: Generalization of the manakov equations. *J. Lightw. Technol.* **2013**, *31*, 398–406. [CrossRef]
25. Ryf, R.; Randel, S.; Gnauck, A.H.; Bolle, C.; Sierra, A.; Mumtaz, S.; Esmaeelpour, M.; Burrows, E.C.; Essiambre, R.-J.; Winzer, P.J.; et al. Mode-division multiplexing over 96 km of few-mode fiber using coherent 6 × 6 MIMO processing. *J. Lightw. Technol.* **2012**, *30*, 521–531. [CrossRef]
26. Carpenter, J.; Wilkinson, T.D. Mode Division Multiplexing over 2 km of OM2 fibre using rotationally optimized mode excitation with fibre coupler demultiplexer. In Proceedings of the IEEE Photonics Society Summer Topical Meeting Series, Seattle, WA, USA, 9–11 July 2012.
27. Weng, Y.; He, X.; Yao, W.; Pacheco, M.; Wang, J.; Pan, Z. Investigation of adaptive filtering and MDL mitigation based on space-time block-coding for spatial division multiplexed coherent receivers. *Opt. Fiber Technol.* **2017**, *36*, 231–236. [CrossRef]
28. Winzer, P.J.; Foschini, G.J. MIMO capacities and outage probabilities in spatially multiplexed optical transport systems. *Opt. Express* **2011**, *19*, 16680–16696. [CrossRef] [PubMed]
29. Weng, Y.; He, X.; Yao, W.; Pacheco, M.C.; Wang, J.; Pan, Z. Mode-dependent loss mitigation scheme for PDM-64QAM few-mode fiber space-division-multiplexing systems via STBC-MIMO equalizer. In Proceedings of the Conference on Lasers and Electro-Optics (CLEO), San Jose, CA, USA, 14–19 May 2017.
30. Raptis, N.; Grivas, E.; Pikasis, E.; Syvridis, D. Space-time block code based MIMO encoding for large core step index plastic optical fiber transmission systems. *Opt. Express* **2011**, *19*, 10336–10350. [CrossRef] [PubMed]
31. Weng, Y.; He, X.; Wang, J.; Pan, Z. Rigorous study of low-complexity adaptive space-time block-coded MIMO receivers in high-speed mode multiplexed fiber-optic transmission links using few-mode fibers. In Proceedings of the SPIE 10130, Next-Generation Optical Communication: Components, Sub-Systems, and Systems VI, San Francisco, CA, USA, 28 January 2017.
32. Mizusawa, Y.; Kitamura, I.; Kumamoto, K.; Zhou, H. Proposal analysis and basic experimental evaluation of MIMO radio over fiber relay system. In Proceedings of the SPIE 10559, Broadband Access Communication Technologies XII, San Francisco, CA, USA, 29 January 2018.
33. He, X.; Weng, Y.; Pan, Z. Fast convergent frequency-domain MIMO equalizer for few-mode fiber communication systems. *Opt. Comm.* **2018**, *409*, 131–136. [CrossRef]
34. Li, G. Recent advances in coherent optical communication. *Adv. Opt. Photonics* **2009**, *1*, 279–307. [CrossRef]
35. Muga, N.J.; Fernandes, G.M.; Ziaie, S.; Ferreira, R.M.; Shahpari, A.; Teixeira, A.L.; Pinto, A.N. Advanced digital signal processing techniques based on Stokes space analysis for high-capacity coherent optical systems. In Proceedings of the 19th International Conference on Transparent Optical Networks (ICTON 2017), Girona, Spain, 2–6 July 2017.
36. Lin, C.; Djordjevic, I.B.; Zou, D. Achievable information rates calculation for optical OFDM few-mode fiber long-haul transmission systems. *Opt. Express* **2015**, *23*, 16846–16856. [CrossRef] [PubMed]
37. Inan, B.; Jansen, S.L.; Spinnler, B.; Ferreira, F.; Borne, D.V.D.; Kuschnerov, M.; Lobato, A.; Adhikari, S.; Sleiffer, V.A.; Hanik, N. DSP requirements for MIMO spatial multiplexed receivers. In Proceedings of the IEEE Photonics Society Summer Topical Meeting Series, Seattle, WA, USA, 9–11 July 2012.
38. Pan, Z.; He, X.; Weng, Y. Frequency domain equalizer in few-mode fiber space-division-multiplexing systems. In Proceedings of the 23rd Wireless and Optical Communication Conference (WOCC), Newark, NJ, USA, 9–10 May 2014.
39. Dick, C.; Amiri, K.; Cavallaro, J.R.; Rao, R. Design and architecture of spatial multiplexing MIMO decoders for FPGAs. In Proceedings of the 42nd Asilomar Conference on Signals, Systems and Computers, Pacific Grove, CA, USA, 26–29 October 2008.
40. Cvijetic, N.; Ip, E.; Prasad, N.; Li, M.J.; Wang, T. Experimental time and frequency domain MIMO channel matrix characterization versus distance for 6 × 28 Gbaud QPSK transmission over 40 × 25 km few mode fiber. In Proceedings of the Optical Fiber Communication Conference (OFC), San Francisco, CA, USA, 9–13 March 2014.
41. Awwad, E.; Othman, G.R.; Jaouën, Y. Space-time coding and optimal scrambling for mode multiplexed optical fiber systems. In Proceedings of the IEEE International Conference on Communications (ICC), London, UK, 8–12 June 2015.

42. Salsi, M.; Koebele, C.; Sperti, D.; Tran, P.; Mardoyan, H.; Brindel, P.; Bigo, S.; Boutin, A.; Verluise, F.; Sillard, P.; et al. Mode-Division Multiplexing of 2 × 100 Gb/s Channels Using an LCOS-Based Spatial Modulator. *J. Lightw. Technol.* **2012**, *30*, 618–623. [CrossRef]
43. Weng, Y.; He, X.; Pan, Z. Space division multiplexing optical communication using few-mode fibers. *Opt. Fiber Technol.* **2017**, *36*, 155–180. [CrossRef]
44. Feuer, M.D.; Nelson, L.E.; Zhou, X.; Woodward, S.L.; Isaac, R.; Zhu, B.; Taunay, T.F.; Fishteyn, M.; Fini, J.F.; Yan, M.F. Demonstration of joint DSP receivers for spatial superchannels. In Proceedings of the IEEE Photonics Society Summer Topical Meeting Series, Seattle, WA, USA, 9–11 July 2012.
45. Feuer, M.D.; Nelson, L.E.; Zhou, X.; Woodward, S.L.; Isaac, R.; Zhu, B.; Taunay, T.F.; Fishteyn, M.; Fini, J.F.; Yan, M.F. Joint digital signal processing receivers for spatial superchannels. *IEEE Photonics Technol. Lett.* **2012**, *24*, 1957–1960. [CrossRef]
46. Ryf, R.; Fontaine, N.K.; Chen, H.; Wittek, S.; Li, J.; Alvarado-Zacarias, J.C.; Amezcua-Correa, R.; Antonio-Lopez, J.E.; Capuzzo, M.; Kopf, R.; et al. Mode-multiplexed transmission over 36 spatial modes of a graded-index multimode fiber. In Proceedings of the European Conference on Optical Communication (ECOC), Rome, Italy, 23–27 September 2018.
47. Ryf, R.; Fontaine, N.K.; Wittek, S.; Choutagunta, K.; Mazur, M.; Chen, H.; Alvarado-Zacarias, J.C.; Amezcua-Correa, R.; Capuzzo, M.; Kopf, R.; et al. High-spectral-efficiency mode-multiplexed transmission over graded-index multimode fiber. In Proceedings of the European Conference on Optical Communication (ECOC), Rome, Italy, 23–27 September 2018.
48. Weerdenburg, J.V.; Ryf, R.; Alvarado-Zacarias, J.C.; Alvarez-Aguirre, R.A.; Fontaine, N.K.; Chen, H.; Amezcua-Correa, R.; Koonen, T.; Okonkwo, C. 138 Tbit/s transmission over 650 km graded-index 6-mode fiber. In Proceedings of the European Conference on Optical Communication (ECOC), Gothenburg, Sweden, 17–21 September 2017.
49. Rademacher, G.; Ryf, R.; Fontaine, N.K.; Chen, H.; Essiambre, R.-J.; Puttnam, B.J.; Luis, R.S.; Awaji, Y.; Wada, N.; Gross, S.; et al. 3500-km mode-multiplexed transmission through a three-mode graded-index few-mode fiber link. In Proceedings of the European Conference on Optical Communication (ECOC), Gothenburg, Sweden, 17–21 September 2017.
50. Rademacher, G.; Ryf, R.; Fontaine, N.K.; Chen, H.; Essiambre, R.-J.; Puttnam, B.J.; Luis, R.S.; Awaji, Y.; Wada, N.; Gross, S.; et al. Long-Haul Transmission Over Few-Mode Fibers with Space-Division Multiplexing. *J. Lightw. Technol.* **2018**, *36*, 1382–1388. [CrossRef]
51. Igarashi, K.; Soma, D.; Wakayama, Y.; Takeshima, K.; Kawaguchi, Y.; Yoshikane, N.; Tsuritani, T.; Morita, I.; Suzuki, M. Ultra-dense spatial-division-multiplexed optical fiber transmission over 6-mode 19-core fibers. *Opt. Express* **2016**, *24*, 10213–10231. [CrossRef]
52. Fontaine, N.K.; Ryf, R.; Chen, H.; Velazquez Benitez, A.; Antonio Lopez, J.E.; Amezcua Correa, R.; Guan, B.; Ercan, B.; Scott, R.P.; Yoo, S.J.B.; et al. 30 × 30 MIMO Transmission over 15 Spatial Modes. In Proceedings of the Optical Fiber Communication Conference (OFC), Los Angeles, CA, USA, 22–26 March 2015.
53. Luís, R.S.; Rademacher, G.; Puttnam, B.J.; Awaji, Y.; Wada, N. Long distance crosstalk-supported transmission using homogeneous multicore fibers and SDM-MIMO demultiplexing. *Opt. Express* **2018**, *26*, 24044–24053. [CrossRef]
54. Garcia-Moll, V.M.; Simarro, M.A.; Martínez-Zaldívar, F.J.; González, A.; Vidala, A.M. Maximum likelihood soft-output detection through Sphere Decoding combined with box optimization. *Signal Process.* **2016**, *125*, 249–260. [CrossRef]

Review

Recent Advances in Equalization Technologies for Short-Reach Optical Links Based on PAM4 Modulation: A Review

Honghang Zhou, Yan Li *, Yuyang Liu, Lei Yue, Chao Gao, Wei Li, Jifang Qiu, Hongxiang Guo, Xiaobin Hong, Yong Zuo and Jian Wu *

The State Key Laboratory of Information Photonics and Optical Communications, Beijing University of Posts and Telecommunications, Beijing 100876, China; zhh1994@bupt.edu.cn (H.Z.); yuyangliu@bupt.edu.cn (Y.L.); leiyuebupt@outlook.com (L.Y.); gcg@bupt.edu.cn (C.G.); w_li@bupt.edu.cn (W.L.); jifangqiu@bupt.edu.cn (J.Q.); hxguo@bupt.edu.cn (H.G.); xbhong@bupt.edu.cn (X.H.); yong_zuo@bupt.edu.cn (Y.Z.)

* Correspondence: liyan1980@bupt.edu.cn (Y.L.); jianwu@bupt.edu.cn (J.W.)

Received: 2 May 2019; Accepted: 3 June 2019; Published: 7 June 2019

Abstract: In recent years, short-reach optical links have attracted much more attention and have come to constitute a key market segment due to the rapid development of data-center applications and cloud services. Four-level pulse amplitude modulation (PAM4) is a promising modulation format to provide both a high data rate and relatively low cost for short-reach optical links. However, the direct detector and low-cost components also pose immense challenges, which are unforeseen in coherent transmission. To compensate for the impairments and to truly meet data rate requirements in a cost-effective manner, various digital signal processing (DSP) technologies have been proposed and investigated for short-reach PAM4 optical links. In this paper, an overview of the latest progress on DSP equalization technologies is provided for short-reach optical links based on PAM4 modulation. We not only introduce the configuration and challenges of the transmission system, but also cover the principles and performance of different equalizers and some improved methods. Moreover, machine learning algorithms are discussed as well to mitigate the nonlinear distortion for next-generation short-reach PAM4 links. Finally, a summary of various equalization technologies is illustrated and our perspective for the future trend is given.

Keywords: short-reach optical links; direct detection; four-level pulse amplitude modulation; digital signal processing; equalization

---

1. Introduction

Driven by upcoming services such as the Internet of Things (IoT), 4K/8K video applications, virtual reality (VR) and big data, the global internet protocol (IP) traffic has grown explosively in recent years [1]. As predicted by a Cisco report [2], 4.8 zettabytes of annual global IP traffic will be reached by 2022 and the so-called "Zettabyte Era" has arrived. To accommodate the associated demands, short-reach optical links have been widely investigated for data center interconnects (DCI), metro network, optical access, and so forth [3]. Unlike long-reach transmission, the short-reach optical links are especially sensitive to cost and size due to the large scale of their deployment [4]. Therefore, intensity modulation with direct detection (IM/DD) is adopted as the mainstream solution instead of coherent detection [5–7]. Traditional IM/DD optical interconnects implemented with non-return-to-zero on-off-keying (NRZ-OOK) format struggle to support the requirements of the increased transmission rate. Thus, many advanced modulation formats are employed to improve the spectrum efficiency (SE) and reduce the bandwidth limitation for electronic and optical components [8–17], such as four-level pulse amplitude modulation (PAM4), carrier-less amplitude and phase modulation (CAP), discrete multi-tone (DMT), and quadrature amplitude modulation (QAM) based on Kramers-Kronig (KK) receiver [18]. Considering the power consumption and implementation complexity, the most attractive format in short-reach optical links is PAM4, which is believed as a highly promising candidate for the next-generation passive optical network (NG-PON) and has been ratified by IEEE 802.3 bs for 400 Gbps Ethernet transmission [19–25].

However, the low-pass filtering effects induced by the limited bandwidth of transmitter and receiver can cause severely inter-symbol interference (ISI). Low-cost devices such as lasers, modulators, photodiodes (PD) and trans-impedance amplifiers (TIA) also produce nonlinear distortions like level-dependent skew and level-dependent noise. Furthermore, the interaction between chromatic dispersion (CD) and direct detection will lead to a power-fading effect, where the detected signal may contain frequency notches after several kilometers transmission at a high symbol rate. Therefore, various equalization technologies based on digital signal processing (DSP) have been investigated for short-reach links over the years [26–53]. Conventional equalizers, such as the feed-forward equalizer (FFE) or decision feedback equalizer (DFE), are employed to compensate for the linear impairments induced by limited bandwidth and CD [26–28], while the equalizer based on the Volterra series is utilized to mitigate nonlinear distortion [29–32]. Some new equalization techniques, such as direct detection-faster than Nyquist (DD-FTN) algorithm [33–35], intensity directed FFE/DFE (ID-FFE/ID-DFE) algorithm [36,37] and joint clock recovery and FFE (CR-FFE) algorithm [38,39], have recently been proposed to solve different problems in practical implementation. In addition, machine learning methods like support vector machine (SVM) [40–42] and neural network (NN) [43–53] have been investigated to further eliminate system impairments for PAM4 modulation.

In this paper, an overview is provided on the important topic of advanced DSP for short-reach PAM4 optical links. The rest of this paper is organized as follows. In Section 2, the system configuration is introduced for short-reach PAM4 transmission in detail, where the comparisons for different transmitters and receivers are presented as well. Section 3 describes the distortions of high-speed transmission with low cost, including limited bandwidth, CD, nonlinear devices and power fading effect. To deal with these impairments in practical implementation, the principles of conventional FFE/DFE and improved algorithms based on FFE/DFE are first illustrated in Section 4. Moreover, VNLE and equalization based on machine learning are also discussed here. Furthermore, a summary of various equalization technologies is given in Section 4. Finally, conclusions are drawn in Section 5.

2. System Configuration

The typical system configuration for short-reach optical links based on PAM4 modulation is presented in Figure 1. The origin data generated by pseudorandom binary sequence (PRBS) is first encoded by a forward error correction (FEC) encoder and then mapped to the PAM4 symbol. After resampling and pulse shaping, the digital signal is pre-equalized through a finite impulse response (FIR) filter to pre-compensate for the limited bandwidth and nonlinearity of the transmitter. The processed signal is then loaded into a digital-to-analog converter (DAC) in order to obtain the baseband electrical signal. Afterward, the DAC output signals are amplified by a linear driver and then fed into a modulator for the generation of optical PAM4 signal. The laser can be a vertical cavity surface-emitting laser (VCSEL), a directly modulated laser (DML) or an electro-absorption modulator integrated with distributed feedback laser (EML). A comparison between different kinds of transmitter is summarized in Table 1. Generally, the directly-modulated lasers like VCSEL and DML reach a bandwidth barrier to operate beyond 25 GBaud [54], despite many research efforts made to achieve higher speed transmission [55,56]. EML generally have high bandwidth at the expense of high cost. Since the electrical signals of VCSEL and DML are directly applied to their laser cavity, larger frequency chirps will exist rather than EML, which can induce severe nonlinear distortions and lead to the serious degradation of transmission performance [57,58]. However, the adiabatic chirp of DML causes frequency modulation (FM) of the optical signal, which is able to fundamentally mitigate the first power fading dip incurred by CD [59]. Note that the Mach-Zehnder modulator (MZM) based on

Lithium Niobate ($LiNbO_3$) is out of consideration due to a large footprint and high cost. For optical links below 300 m, the multi-mode fiber (MMF) combined with VSCEL is widely used at present. When the transmission distance increases, the modal dispersion in the MMF will distort the signal and the standard single-mode fiber (SSMF) becomes the common choice.

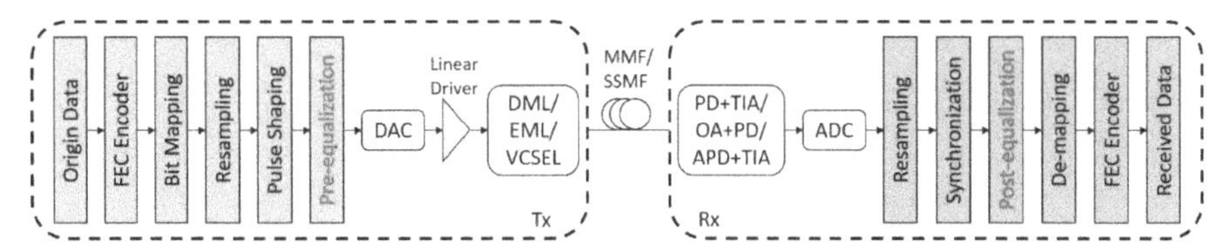

Figure 1. Typical configuration for short-reach PAM4 transmission system.

Table 1. Comparison of different transmitters [54–59].

Tx	Bandwidth	Chirp	Cost	Reach	Fiber	Power Fading	Wavelength (nm)
VSCEL	low	high	low	<300 m	MMF	N/A	Mostly 850
DML	low	high	fair	<80 km	SSMF	low	1310/1550
EML	high	low	high	<80 km	SSMF	high	1310/1550

At the receiver, the signal is directly detected by PD or avalanche photodiode (APD). To improve receiver sensitivity, a TIA is generally cascaded behind to amplify the electrical signal. An alternative solution is using the combination of an optical amplifier (OA) and an optical band-pass filter (OPBF) before PD to amplify received optical signal and remove out-of-band noise. The comparison of different receivers is presented in Table 2, where OA + PD has the highest sensitivity due to the use of high-cost OA. The gain of APD results in relative higher SNR, which makes it more suitable for longer reach links compared to traditional PD [26]. After being processed by an analog-to-digital converter (ADC), the signal is resampled to an expected sampling rate for subsequent steps. Synchronization is achieved to eliminate sampling clock offset (SCO) and depress timing jitter. Then, various DSP techniques are utilized to equalize the received signal for the performance improvement of the transmission system. Finally, after de-mapping and FEC decoding, the received data is obtained and the bit error rate (BER) can be calculated.

Table 2. Comparison of different receivers [1,7,26,39].

Rx	Cost	Footprint	Sensitivity	Reach
PD + TIA	low	low	low	low
APD + TIA	fair	low	fair	fair
OA + PD	high	high	high	high

However, due to the limitation of cost, all transmitters and receivers mentioned above can induce different degree distortions for high-speed PAM4 transmission. Therefore, equalization is one of the most DSP components to eliminate impairments in short-reach optical links. As is shown in Figure 1, the equalization technologies can be utilized both in the transmitter as pre-equalization and in the receiver as post-equalization. The distortions induced by low cost and the different equalization technologies will be introduced specifically in the following sections.

3. Distortions Induced by Low Cost

In short-reach optical links based on PAM4 modulation, the cost is one of the most important factors to be considered for commercial implementation. Low-cost components are highly desired in practical application, while they also bring great challenges and serious distortions such as linear and nonlinear impairments. In this section, the major distortions induced by low cost will be illustrated in short-reach optical communications.

3.1. Linear Impairments: Limited Bandwidth and Chromatic Dispersion

The high-speed PAM4 system suffers from severe bandwidth limitation due to the consideration of low cost. After bandwidth limited devices, the received signal with low-pass filtering effects can be expressed as:

$$y_k = I_k + \sum_{\substack{n=0 \\ n \neq k}}^{\infty} I_n x_{k-n} + v_k \tag{1}$$

where x_k and y_k is the input signal and output signal at the k-th sampling instant, respectively. The first term I_k represents the desired information symbol while the second term donates the ISI. The last term v_k is the additive Gaussian noise variable at the k-th sampling instant. To investigate the performance impact of bandwidth limitation, a typical measured system frequency response is shown in Figure 2a [60]. The 3 dB bandwidth of the total transmission system is approximately 7.5 GHz, which is far below the required bandwidth for 64 Gbaud PAM4 signal. The attenuation of amplitude frequency is about 15 dB at the Nyquist frequency. In the insets (I) of Figure 3a, it can be observed that the optical eye diagram is confusing, and even four levels of PAM4 signal are unable to be detected. After compensation, a great improvement is achieved in the quality of the eye diagram, which reveals the tremendous influence of limited bandwidth. As shown in the blue part of Figure 2b, the receiver sensitivity penalty is measured as a function of the bandwidth for PAM4 system in simulation [19]. The received optical power at BER = 3.8×10^{-3} for a bandwidth of 35 GHz is employed as reference. It can be seen that approximately 3.3 dB received power is lost as the bandwidth decreases from 35 GHz to 15 GHz. Moreover, the receiver sensitivity penalty increases in speed when the bandwidth limitation becomes severe. The other simulation for PAM4 transmission is depicted in the black part of Figure 2b, where the bandwidths of the components are varied uniformly [61]. The total bit rate is fixed at 96 Gbps, while the 3 dB bandwidths of DAC, modulator, PD and ADC are changed from −30 to 30% simultaneously. The obtained BER performance is degraded as the bandwidth reduces, which confirms that the narrow bandwidth is one of the factors that deteriorate the performance of PAM4 optical links.

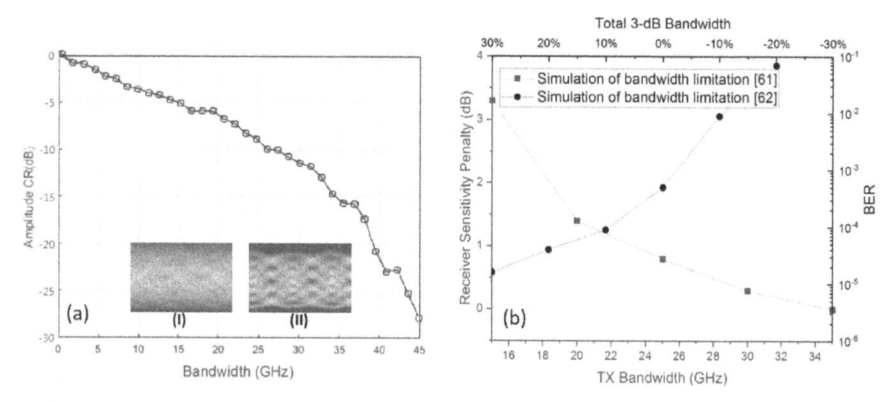

Figure 2. (**a**) Frequency response in band-limited system and eye diagrams (I) before and (II) after bandwidth compensation (redrawn after [60]). (**b**) Effect of bandwidth to system performance (redrawn after [19,61]).

When PAM4 is used in the C-band, the CD tolerance is considered as one key drawback, which critically impairs the high-frequency components of the signal. The typical tolerance to residual

CD in terms of the optical signal-to-noise ratio (OSNR) for PAM4 is shown in Figure 3, where the OSNR penalty is about 1 dB for dispersion within ±100 ps/nm [62]. Meanwhile, the eye diagrams for 56 Gbps PAM4 signal after 0 km and 20 km transmission are shown in Figure 3b,c, respectively. A qualitative comparison between these two eye diagrams reveals that CD brings serious degradation in signal quality.

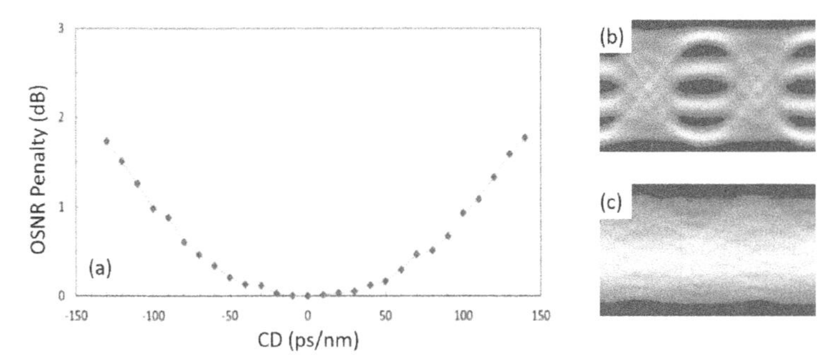

Figure 3. (**a**) Four-level pulse amplitude modulation (PAM4) signaling tolerance on dispersion in terms of optical signal-to-noise ratio (OSNR) penalty (reprinted from [62] with permission from authors). Eye diagrams for 56 Gbps PAM4 signal after (**b**) 0 km and (**c**) 20 km transmission.

3.2. Nonlinear Impairments

3.2.1. Nonlinear Devices: Level-Dependent Noise and Level-Dependent Skew

The low cost of devices not only causes the bandwidth limitation but also produces nonlinear distortions like amplitude-dependent noise and level-dependent skew. An example for APD or PD with the optical amplifier is depicted in Figure 4, where different power levels of the PAM4 signal have different distributions of dominant noise [63]. The probability of distributions of regular equally-spaced PAM4 signal after OA + PD and APD receivers are illustrated in Figure 4a,b, respectively. The peaks of probability density represent different noise variances for different levels. From Figure 4c we can notice that high power levels suffer larger noise interference and more symbols cross the decision thresholds, which is consistent with the probability of distributions. Note that the distribution of noise depends on which impairment is dominant, so it is diversified. For instance, the noise of two middle levels could be larger than the noise of two marginal levels considering the nonlinearity of the modulation curve.

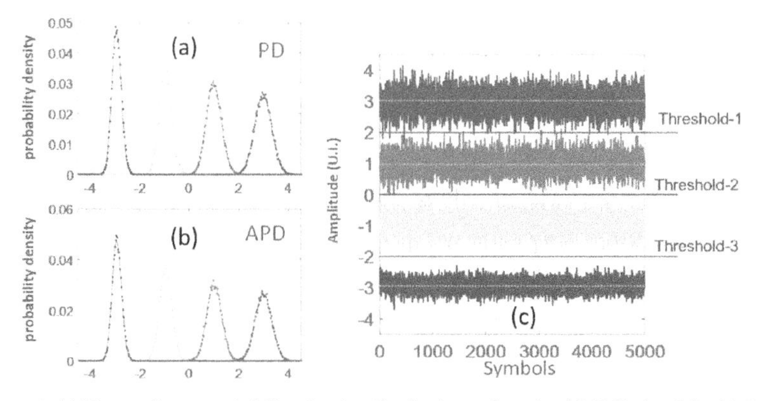

Figure 4. (**a**) The nonlinear probability density distributions of received PAM4 signal for (**a**) OA + PD, and (**b**) APD. (**c**) The waveform of received PAM4 signal after APD (reprinted from [63] with permission from authors).

The level-dependent skew induced by the nonlinear chirp characteristic of modulators can produce significant penalties due to the differences in optimum sampling time for the different level of the eye. As shown in Figure 5a,b, the simulated PAM4 signal without any skew and the simulated 50 Gb/s PAM4 signal using spatial-temporal rate equation modeling [64]. An obvious amplitude-dependent eye can be observed, and hence this inevitably introduces timing impairments. The measured eye diagrams for different signal rates and bias currents of the laser are presented in Figure 5c–f. These figures indicate that the phenomenon of level-dependent skew is more serious under the condition of high-speed and low power consumption, which severely degrades the transmission performance.

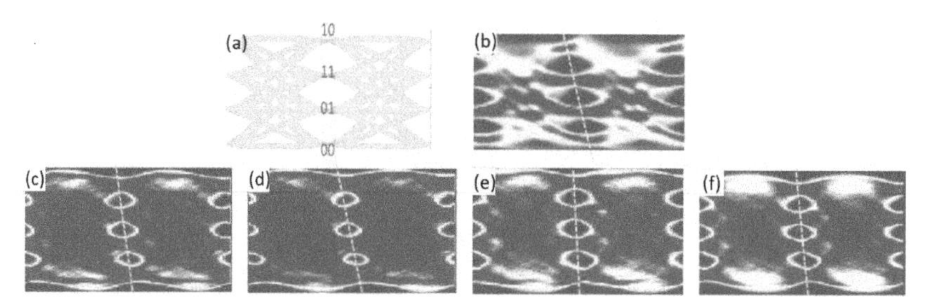

Figure 5. The eye diagrams of (**a**) ideal PAM4 signal; (**b**) simulated PAM4 at 50 Gb/s using spatial-temporal modeling, (**c**) measured 56 Gb/s PAM4 signal when bias current is 5 mA; (**d**) measured 64 Gb/s PAM4 signal when bias current is 5 mA; (**e**) measured 56 Gb/s PAM4 signal when bias current is 7 mA; (**f**) measured 64 Gb/s PAM4 signal when bias current is 7 mA (reprinted from [64] with permission from authors).

3.2.2. Power Fading Effect

Due to the interaction between chromatic CD and direct detection, the induced power-fading effect will produce spectral zeros in the spectrum of the signal, which is the key points to determine transmission distance. Figure 6a–c plots the simulated magnitude response of a 28 Gbaud PAM4 system for various fiber lengths, namely 15 km, 50 km, and 100 km, respectively [65]. It can be clearly seen that the number of spectral zeros changes from none to one and then to three by increasing the transmission distance. Meanwhile, the probability of frequency notches in the high frequency is greater, which proves that the power-fading effect is more severe for high-speed transmission. For experimental demonstrate, the frequency response of the received 25 Gbaud PAM4 signal over 50 km SSMF transmission is depicted in Figure 6d [66]. The first frequency notch can be obviously found around 9 GHz, which indicates that the power fading effect is an inescapable problem in IM/DD system.

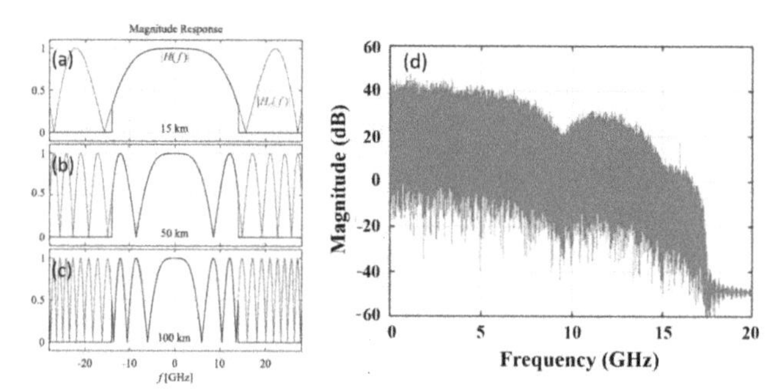

Figure 6. Magnitude responses of a 28 Gbaud PAM4 system after (**a**) 15 km; (**b**) 50 km; (**c**) 100 km in simulation (reprinted from [65] with permission from IEEE). (**d**) Frequency spectrum of the detected 25 Gbaud PAM4 signal over 50 km SSMF transmission (reprinted from [66] with permission from authors).

4. Equalization Technologies

To mitigate the impairments in high-speed PAM4 system mentioned above, various digital equalization technologies have been studied. In this section, we introduce the most popular equalizers, including FFE, DFE, Volterra nonlinear equalizer (VNLE) and machine learning based equalizer, in detail to compensate for these impairments. Moreover, some improved equalization methods are also described to handle specific issues here.

4.1. FFE/DFE

4.1.1. Conventional FFE/DFE

The feed-forward equalizer is an effective method for linear impairments compensation and widely used nowadays. The most basic components of FFE is the finite impulse response (FIR) filter, whose structure is shown in Figure 7a. The output of FIR is expressed as [67]

$$y(k) = \sum_{l=0}^{n-1} h_l x(k-l) \tag{2}$$

where $x(k)$ and $y(k)$ are the input and output signal of FIR at the sampling instant k, respectively. $h = [h_0 h_1 h_2 \ldots h_{n-1}]$ is the array of tap weights, while n is the number of taps.

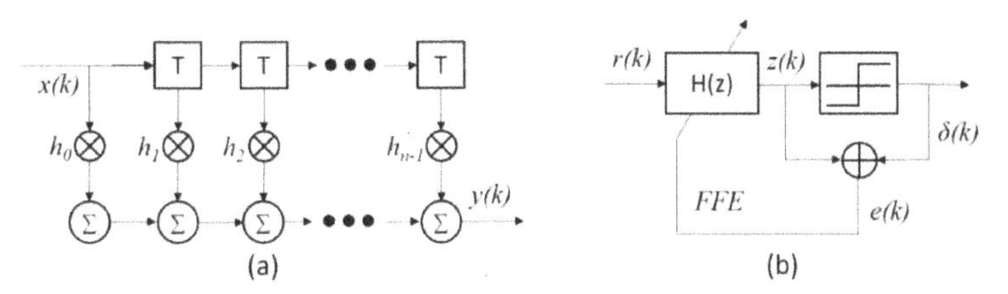

Figure 7. The structure of (**a**) FIR filter: H(z); (**b**) FFE.

Figure 7b depicts the structure of decision-directed FFE, where the FIR filter is noted as $H(z)$. The tap weights can be updated by the zero-forcing (ZF) algorithm, least mean squares (LMS) algorithm, recursive least squares (RLS) algorithm and so on [67–69]. Different convergence algorithms just affect the speed of obtaining optimal tap weights and are not the focus of this review. Here, we only illustrate one of the most common techniques called direct decision LMS (DD-LMS) algorithm. The $(k+1)$-th update of the filter tap weights is given by [68]

$$h(k+1) = h(k) + 2\mu e(k)R(k) \tag{3}$$

where μ is the step size and the error $e(k) = \delta(k) - z(k)$ is between the desired signal $\delta(k)$ and the output signal $z(k)$. The $R(k) = [r(k), r(k-1), r(k-2), \ldots, r(k-n+1)]$ is the vector of the input signal. The FFE can boost the power of high-frequency components that undergo large losses due to the system bandwidth limitations. It should be noted that FFE can operate at symbol rate sampling or higher. Compared with symbol-spaced FFE, the fractionally-spaced FFE which is sampled at several times the symbol rate allows the matched filter to be realized digitally and can take care of phase recovery. In addition, it has a lower residual error at the cost of computational complexity. The destructive effect of the frequency notches is unable to be compensated for easily with an FFE but could be efficiently mitigated using a DFE. Unlike FFE, the input of DFE is the signal after decision, as shown in Figure 8a.

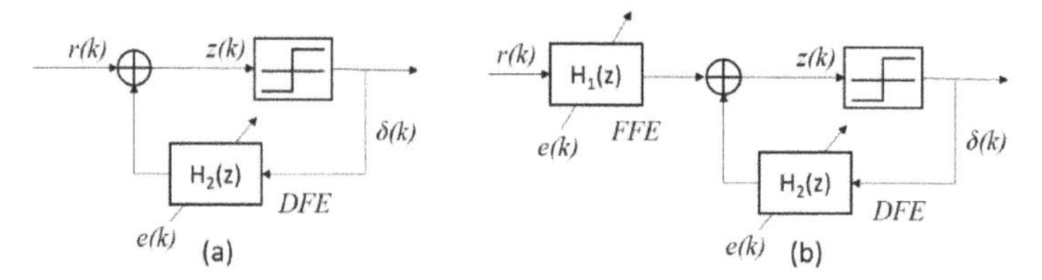

Figure 8. The structure of (**a**) DFE; (**b**) FFE and DFE combined.

In the case of DFE, the signal before decision is calculated by [69]

$$z(k) = r(k) - \sum_{l=0}^{n-1} h_l \delta(k-l) \tag{4}$$

The $(k+1)$th update of the filer tap weights is given by [69]

$$h(k+1) = h(k) + 2\mu e(k) Z(k) \tag{5}$$

where $Z(k) = [z(k), z(k-1), z(k-2), \ldots, z(k-n+1)]$. DFE is usually operated at one sample per symbol. It should be noted that DFE can successfully equalize frequency notches by pole insertion, but it may suffer from error propagation and is unstable due to the decision feedback scheme. Moreover, only post-cursor ISI can be deal with DFE, while the CD-induced channel impulse response contains both pre-cursor and post-cursor. Therefore, the best choice for practical implementation is a combination of an FFE and a DFE, which can be seen from Figure 8b. The FFE and DFE can be placed in the transmitter for pre-compensation or in the receiver for post-compensation [60]. To solve the problem of error propagation, a transmitter-side DFE called Tomlinson-Harashima pre-coding has recently been proposed and investigated in various PAM4 short-reach optical links [70–73]. In addition, many novel equalizers based on conventional FFE/DFE are proposed and investigated to deal with the problems in practical implementation and improve the transmission performance. Next, three recently proposed equalizers including DD-FTN, ID-FFE/ID-DFE, and CR-FFE will be illustrated as examples.

4.1.2. Improved Algorithms Based on FFE/DFE

• DD-FTN

When the FFE tries to amplify the power of high-frequency components to compensate for the low-pass effects, it boosts the noise power at high-frequency components as well. To address this phenomenon, a DD-FTN algorithm is proposed [33] and the principle is shown in Figure 9.

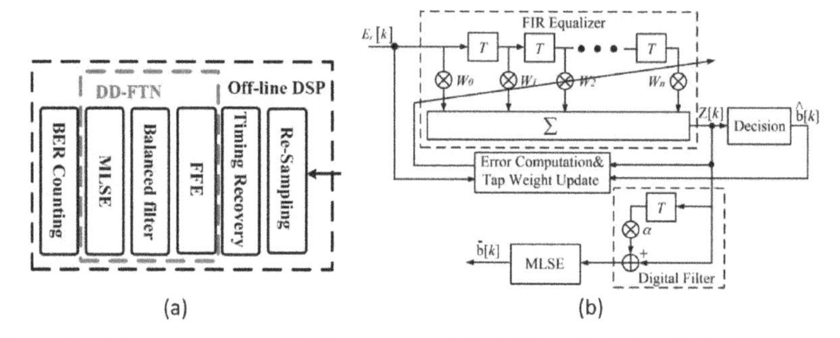

Figure 9. (**a**) The components and (**b**) the structure of DD-FTN (reprinted from [33] with permission from authors and reprinted from [7] with permission from IEEE).

First, the signal after synchronization is equalized by a FFE, adapted by the LMS algorithm. Then, a balanced digital filter is placed behind to suppress the enhanced in-band noise by the equalizer. The transfer function of the balanced filter in z-transform is $H(z) = 1 + \alpha z^{-1}$, where the tap coefficient α is employed to optimize the frequency response of the post filter. Finally, the maximum likelihood sequence estimation (MLSE) is utilized behind to eliminate the strong ISI induced by the balanced filter. Although two more DSP blocks are contained in DD-FTN compared to traditional equalizers, the increase in complexity is relatively small and the overall complexity is not high compared to other DSP techniques.

The BER performance as a function of received optical power for 112 Gbps PAM4 signal after 2 km SSMF transmission is depicted in Figure 10a [33]. With conventional FFE, the enhanced noise significantly degrades the system performance and a BER floor at 5×10^{-2} can be observed. By using DD-FTN, the noise enhancement is effectively mitigated by the balanced filter and the induced ISI can be processed by the MLSE. The performance is improved and the BER can reach 3×10^{-4} when received power is −7.1 dBm. Figure 10b shows the BER performance versus received optical power for 140 Gbit/s PAM-4 system after back-to-back transmission [35]. It can be found that the traditional FFE updated by DD-LMS exhibits the worst performance, while the BER can reach the hard decision FEC (HD-FEC) threshold of 3.8×10^{-3} with DD-FTN. Therefore, DD-FTN can make a big difference in enhanced noise mitigation.

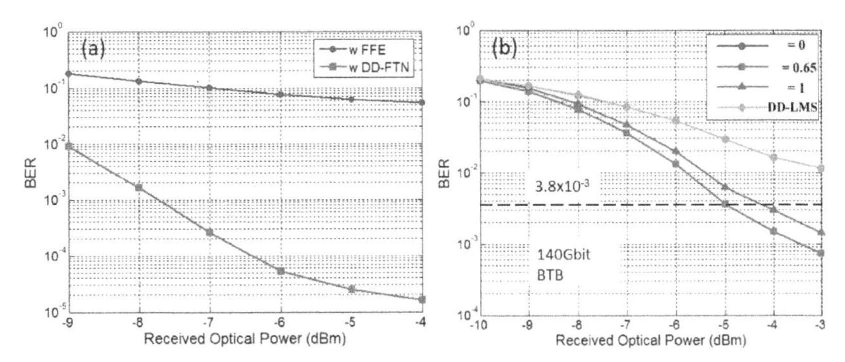

Figure 10. (**a**) The bit error rate (BER) performance versus received optical power for 112 Gbps PAM4 signal after 2 km SSMF transmission (reprinted from [33] with permission from authors). (**b**) The BER performance versus received optical power for different tap coefficient of balanced filter for 140 Gbit/s PAM4 system after back-to-back transmission (reprinted from [35] with permission from IEEE).

- ID-FFE/ID-DFE

In C-band transmission, low-cost modulator like DML will introduce an additional impairment due to the interaction between frequency chirp and chromatic dispersion. Recently, a low complexity intensity directed equalizer based on FFE and DFE is proposed to suppress the chirp induced distortions. Different from the traditional FFE/DFE that utilizes one set of coefficients, the proposed algorithm first divides the symbol into different sets according to its intensity level, as shown in Figure 11a [37]. Then the coefficients of different groups are applied accordingly. Four sets of coefficients for PAM4 symbol are selected according to the three thresholds for ID-FFE and four levels for ID-DFE. After classification, the desired tap coefficients from different sets are used and the following procedures are the same as the traditional FFE/DFE. In terms of complexity, only a few simple decision circuits are added to the original FFE/DFE and the complexity increment is negligible considering the operations of multiplication in the equalizer.

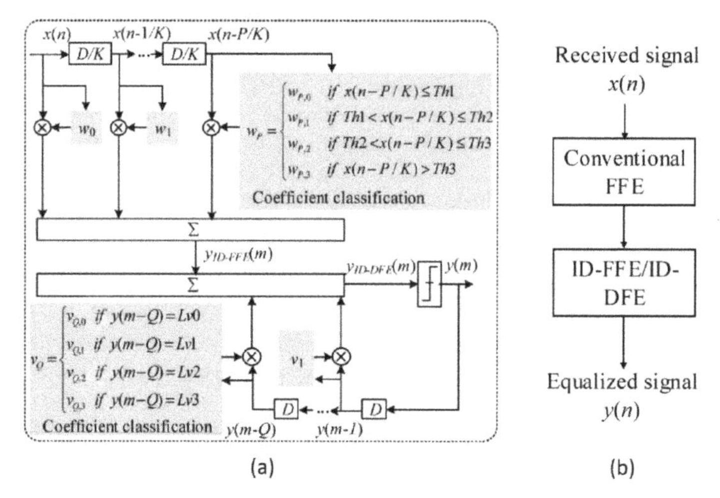

Figure 11. (**a**) The detailed structure of (P-1)-tap ID-FFE and Q-tap ID-DFE. (**b**) The block diagram of the pre-FFE + ID-FFE/ID-DFE (reprinted from [37] with open access from OSA).

The effectiveness of ID-FFE/ID-DFE greatly depends on the accuracy of the decision in the classification. Thus, a conventional FFE is employed to equalize the severe impairments and enhance the decision accuracy before the intensity directed equalizer as shown in Figure 11b.

The BER performance versus the received optical power is shown in Figure 12 [37]. Compared with the FFE case, of which the BER is 9×10^{-3}, the ID-FFE/ID-DFE reaches 2×10^{-3}. When the pre FFE is used before ID-FFE/ID-DFE, the performance is improved by almost an order of magnitude and the BER reduce to 2.6×10^{-4}. From the received eye diagrams of ID-FFE with and without pre-FFE, we observe that the pre-FFE can completely suppress the residual eye skewing effect of the ID-FFE and dramatically remove the ISI. When the transmission distance increase to 43 km, only the ID-FFE/ID-DFE assisted by pre-FFE can reach a BER of 3.6×10^{-3} below the HD-FEC. Therefore, the chirp of the modulator can be well addressed by the novel proposed ID-FFE/ID-DFE.

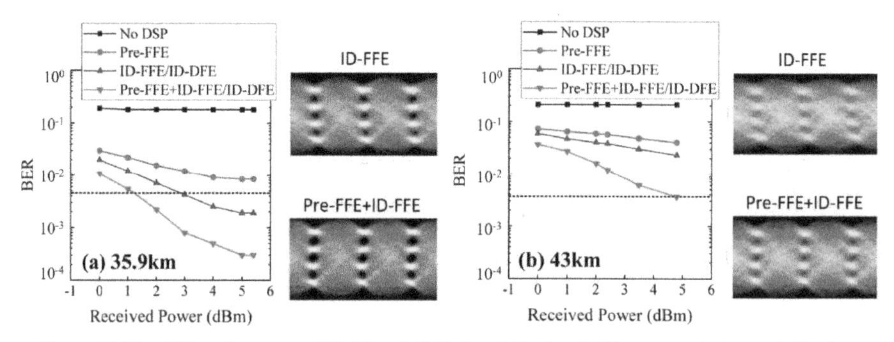

Figure 12. The BER performance of 56 Gbps PAM4 signal (**a**) after 35.9 km transmission and (**b**) after 43 km transmission (reprinted from [37] with open access from OSA).

• CR-FFE

The problem of incompatible prerequisites between impairment equalization and clock recovery also reduces the performance of PAM4 transmission [74]. A joint clock recovery and feed-forward equalization algorithm is proposed recently, which estimates timing error based on the difference between two tap coefficients of T/2-spaced FFE. The proposed algorithm can eliminate the ISI induced by linear impairments and track large sampling clock offset (SCO) simultaneously.

The structure of the CR-FFE is plotted in Figure 13, where we notice that the timing error is derived from the difference between two tap coefficients [39]. For the comparison of complexity, the timing error of CR-FFE is calculated by one time subtraction, while the conventional CR algorithm needs an additional multiplication. Considering the total multiplication operations, the computational complexity of CR-FFE is similar to or slightly lower than that of the original scheme.

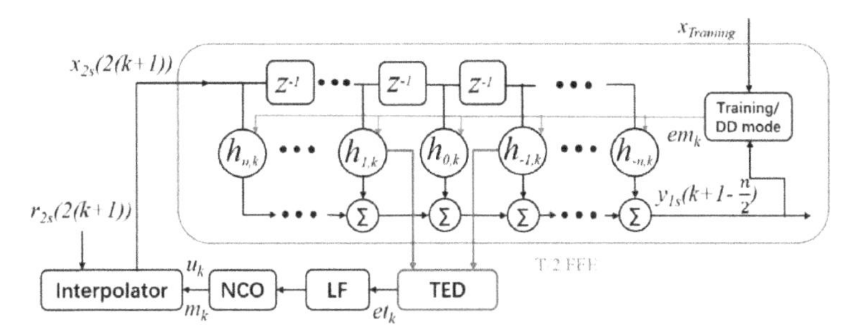

Figure 13. The structure of proposed joint clock recovery and FFE (CR-FFE) (reprinted from [39] with open access from OSA).

The BER performance of CR-FFE over different SCO is studied as shown in Figure 14a [39]. The traditional clock recovery cascaded by FFE (noted as scheme II) cannot counteract large SCO, while the CR-FFE (noted as scheme III) can ensure stable and reliable performance as SCO increase from 0 to 1000 ppm. From Figure 14b,c, we observe that the two tap coefficients for timing error detection are basically equal and the fractional interval has a stable changing process, which indicates that clock recovery and equalization are accomplished simultaneously after 40 km transmission when SCO reaches 1000 ppm. Therefore, the CR-FFE can solve the problem of incompatible prerequisites between impairment equalization and clock recovery, and provide significant system performance improvement for PAM4 transmission.

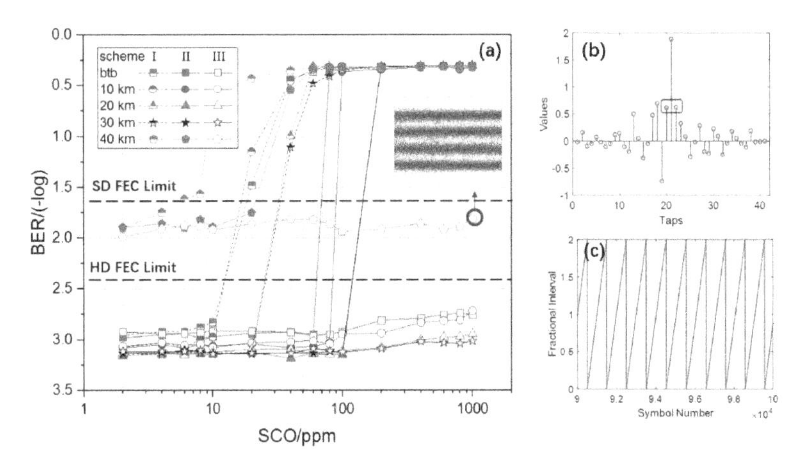

Figure 14. (**a**) The BER performance vs. sampling clock offset at received optical power of −8 dBm for 50 Gbps PAM4 signal; (**b**) the tap weights and (**c**) fractional interval of proposed CR-FFE after 40 km transmission when sampling clock offset is 1000 ppm (reprinted from [39] with open access from OSA).

4.2. VNLE

Thanks to FFE and DFE, the linear impairments can be efficiently eliminated. However, the residual nonlinear distortion mainly induced by the nonlinearity of devices and square-law detection can also severely impact the transmission performance of the PAM4 system. One of the most popular equalization techniques is VNLE, whose structure is shown in Figure 15. Note that the VNLE can be implemented based on FFE or DFE and we focus on FFE based VNLE in this paper. Considering the exponential growth of computational complexity, the third-order VNLE is introduced here.

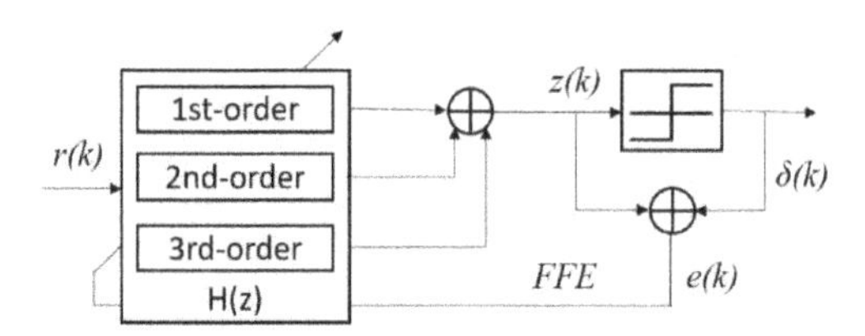

Figure 15. The structure of Volterra nonlinear equalizer (VNLE).

The output of VNLE is expressed as [75]

$$z(k) = \sum_{l=0}^{n_1-1} h_l x(k-l) + \sum_{l=0}^{n_2-1}\sum_{i=0}^{l} h_{l,i} x(k-l)x(k-i) + \sum_{l=0}^{n_3-1}\sum_{i=0}^{l}\sum_{j=0}^{i} h_{l,i,j} x(k-l)x(k-i)x(k-j) \quad (6)$$

where h_l, $h_{l,i}$ and $h_{l,i,j}$ are the tap weights of 1st-order, 2nd-order and 3rd-order kernels, respectively. $x(k)$ is the input signal of FIR in VNLE at the sampling instant k. In addition, n_1, n_2 and n_3 respectively represent the number of taps for the linear part, 2nd-order nonlinear part and 3rd-order nonlinear part. Generally, the linear impairments of the PAM4 system and the self-phase modulation (SPM) of SSMF can be successfully eliminated by the first and third-order kernels of VF, while the second-order kernels are utilized to compensate the nonlinearity of devices such as modulator and PD. The kernel coefficients are commonly updated by LMS algorithm considering complexity. To simplify the convergence procedure, the kernels of each order are evolved at different speeds [75], which can be depicted as a gradient vector $\mu = [\mu_1, \mu_2, \mu_3]$. However, there is no clear procedure on how to set the values of this vector.

While limited by the high computation complexity of fully-connected VNLE, it is not practical to be implemented directly in the real applications. Various methods have been proposed to reduce the computational complexity of VNLE and the interested readers can refer to [76–80] for a detailed discussion on this topic. Taking modified Gram-Schmidt orthogonal decomposition as an example, the performance of sparse VNLE is described as shown in Figure 16 [31]. The dependence of the BER on the back-to-back system without VNLE, and with VNLE or sparse VNLE are compared. As is shown, the required ROP at the FEC threshold is decreased by 0.7 dB for the VNLE. When sparse VNLE is employed, the increase in the ROP is less than 0.2 dB compared to VNLE. Figure 6b plots the dependence of the BER on the ROP for 10 km and 20 km transmission. No penalties are observed relative to the back-to-back system. Therefore, the sparse VNLE can maintain basically the same performance with a reduction of computational complexity, which is an excellent optimization for VNLE.

4.3. Equalization Based on Machine Learning

Machine learning (ML) is the scientific study of algorithms and statistical models that computer systems use to effectively perform a specific task without using explicit instructions, which was coined in 1959 by Arthur Samuel [81]. Recently, machine learning algorithms have been utilized to process optical communications and achieve distinguished performance [82–86]. Some DSP techniques based on machine learning including SVM and NN are described in this section for PAM4 optical links.

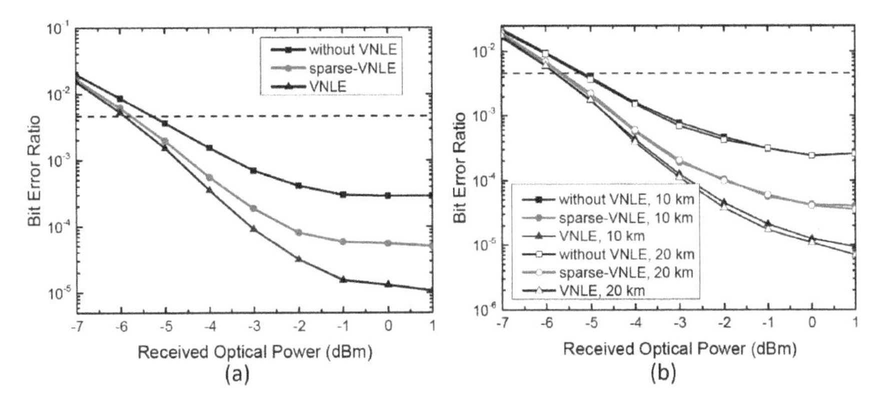

Figure 16. (**a**) BER vs. received optical power for a back-to-back system. (**b**) BER vs. received optical power for 10 km and 20 km transmission (reprinted from [31] with permission from authors).

4.3.1. Support Vector Machine

The basic SVM classifier is a two-class classifier and its training process can be described as finding the maximum edge hyperplane to define the decision function of the classifier. The concept of basic SVM classifier is shown in Figure 17a, where the margin means the minimum distance of all samples to the hyper-plane [40]. To solve this convex quadratic programming of optimization target in SVM, the sequential minimal optimization (SMO) algorithm [87] can be used.

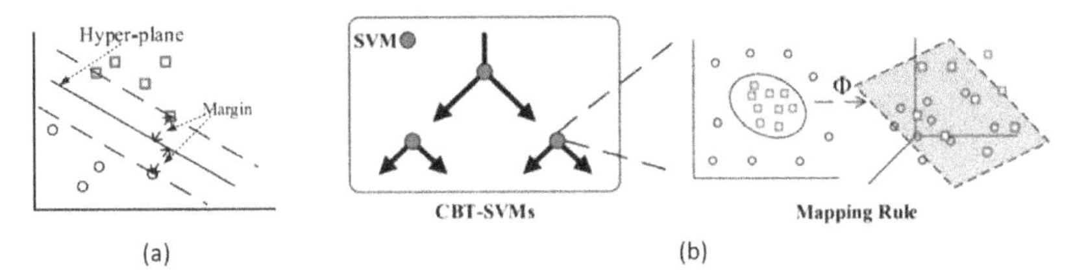

Figure 17. (**a**) Basic SVM classifier. (**b**) Complete binary tree structure multi-classes SVMs and mapping rule for PAM4 signal where the black circle and the red square represent two categories (reprinted from [40] with permission from IEEE).

Note that PAM4 cannot be de detected directly by the basic SVM classifier. Thus, a proposed complete binary tree (CBT) structure multi-classes SVMs method is utilized as depicted in Figure 17b. It can be seen that two layers are contained in the tree, where the first layer classifies the first bit of PAM4 signal, and the second layer classifies the last bit. Although the wrong decision in the high layer will be retained to the next layers, a significant improvement in performance for PAM4 system can still be expected.

The BER curves of the 40 Gbps and 50 Gbps PAM4 signal are shown in Figure 18a,b, respectively [42]. Using the proposed CBT-SVMs, significant receiver sensitivity reduction of 2.99 dB and 4.68 dB in optical B2B and 2 km SMF transmission can be obtained for 40 Gbps PAM4 system. Moreover, for 50 Gbps PAM4 modulation, it provides 3.59 dB and 3.63 dB received power tolerance compared with the hard decision. Therefore, the SVM algorithm shows outstanding performance in nonlinear impairment mitigation and could be a potential choice for future PAM4 optical links.

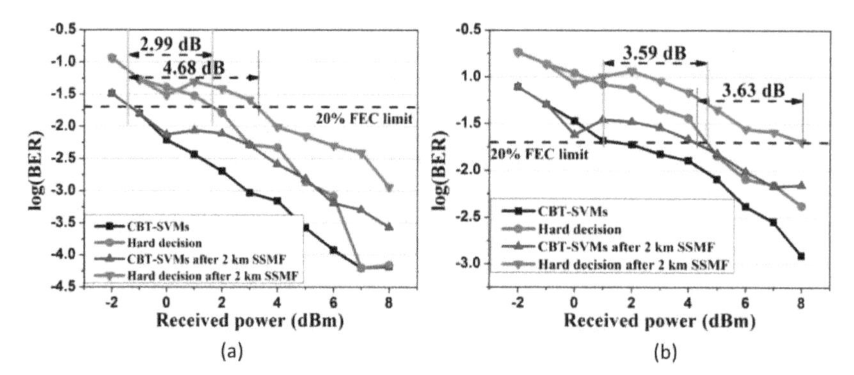

Figure 18. BER curves for (**a**) 40 Gbps and (**b**) 50 Gbps PAM4 signal using CBT-SVMs (reprinted from [42] with permission from IEEE).

4.3.2. Neural Network

The neural network (NN) is computational model loosely inspired by its biological counterparts [88]. In recent years, it has been proposed to mitigate the nonlinear impairments in optical communication system [89–91]. For short-reach PAM4 optical links, various research concerning the NN method has been performed to improve transmission performance [43–53]. The schematic of NN based nonlinear signal processing is presented in Figure 19a, where the leftmost part consists of a set of neurons representing the input features and the rightmost part is a non-linear activation function [48]. In the middle, a weighted linear summation is employed to represent connections between neurons. Figure 19b shows a two-layer neural network to classify PAM4 signals. Rectified Linear Units (ReLU), as indicated in the inset of Figure 19b, is always applied as the activation function. Compared with sigmoid function or Tanh function, it is more like a real neuron in our body and results in much faster training. A softmax function is commonly selected as the activation function for the output layer. The output of the softmax function is the class with the highest probability.

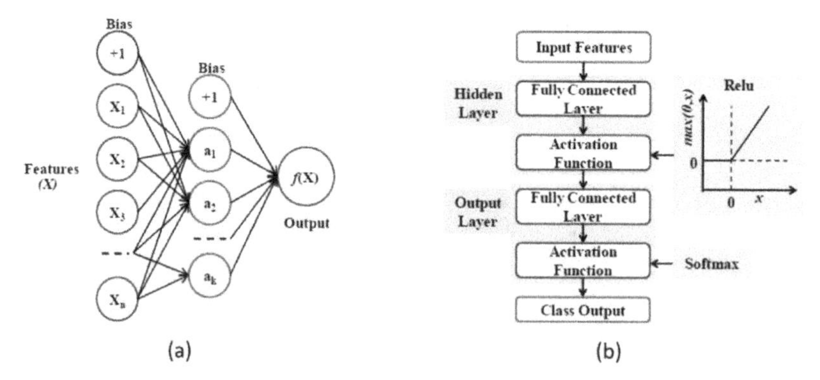

Figure 19. (**a**) Schematic of neural network (NN) based nonlinear signal processing for a hidden layer and (**b**) simple structure of two-layer neural network (reprinted from [48] with permission from authors).

An example for BER performance of NN is shown in Figure 20a [49]. Thanks to the NN method, about one order of magnitude is decrease compared to the BER with VNLE after 60 km transmission. Figure 20b shows the BER performance versus OSNR after 80 km SSMF. The required OSNRs to reach the FEC threshold for VNLE and NN are 39 dB and 35.5 dB, respectively. Thus, we can conclude that compared to conventional equalizers, a better BER performance can be achieved using the NN method, which is an attractive solution for short-reach PAM4 optical links.

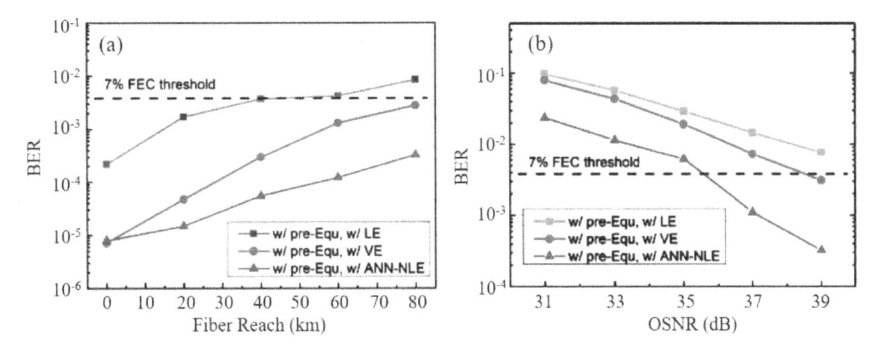

Figure 20. (**a**) BER vs. fiber length with NN for 112 Gbps SSB-PAM4 transmission. (**b**) BER vs. OSNR after 80 km dispersion uncompensated SSMF (reprinted from [49] with open access from OSA).

4.4. Summary of Recent Work

Various equalization technologies mentioned above are summarized and compared as shown in Table 3. The FFE/DFE is widely used in short-reach PAM4 optical links due to simple architecture and low complexity, however, it can only eliminate the linear ISI induced by bandwidth limitation and CD. In [26], a single-channel 56 Gbit/s PAM4 optical transmission over 60 km is experimentally demonstrated with receiver sensitivities of −19.9 dBm using 35 taps FFE. Meanwhile, with the help of 17 taps FFE, the 40 km error-free transmission is achieved for 106 Gbit/s PAM4 using APD receiver [27]. The THP is a transmitter-side DFE without suffering from error-propagation, which is proposed recently to resist against power fading and bandwidth limitation. For the first time, the THP is applied to 100 Gbit/s FTN PAM4 transmission over 40 km through a 20 GHz low-pass channel [72]. With similar or a low addition of complexity, the novel DD-FTN, ID-FFE/ID-DFE and CR-FFE can compensate for the weakness of conventional linear equalizers and achieve better performance for specific issues. In [35], the DD-FTN is experimentally demonstrated for a 140 Gbit/s PAM4 signal over 20 km transmission with a receiver sensitivity of −5.5 dBm. K. Zhang et al. firstly demonstrate a C-band 56 Gbit/s PAM4 system over 43 km transmission with the proposed ID-FFE/ID-DFE [37]. In [39], the CR-FFE resist SCO up to 1000-ppm after 40 km transmission for 50 Gbit PAM4 system based on 10 GHz DML. While bandwidth limitation can be efficiently compensated for by using FFE/DFE, the nonlinearity of devices is difficult to eliminate effectively. VNLE is the most popular nonlinear equalization while the high computational complexity can be reduced by removing the less important kernels. Using 1722 taps VNLE, the C-band 2 × 56 Gbit/s PAM4 transmission over 100 km is experimentally demonstrated in [92]. A sparse VNLE with half computational complexity achieves similar performance for 2×64 Gbit/s PAM4 transmission over 70 km SSMF using 18 GHz DML [93]. As for machine learning algorithms, SVM and NN including artificial NN (ANN), deep NN (DNN), and convolutional NN (CNN) are the mainstream to further eliminate nonlinear distortions of PAM4 optical links with higher complexity. When the SVM is applied to PAM4 signal, sensitivity gain of 2.5 dB is obtained for 60 Gbit/s VCSEL-MMF short-reach optical links [40]. Yang et al. experimentally demonstrate a C-band 4 × 50 Gbit/s PAM4 transmission over 80 km SSMF employing a radial basis function ANN [44]. In [46], with the help of CNN, a 112 Gbit/s PAM4 transmission over 40 km SSMF is accomplished and the BER performance outperforms traditional VNLE. For more complicated DNN, the BER of 4.41×10^{-5} is obtained for a 64 Gbit/s PAM4 transmission over 4 km MMF based on 850 nm VCSEL [47]. Note that some equalization methods such as look-up table (LUT) are not mentioned in this paper due to less implementation.

Table 3. Equalization technologies for short-reach PAM4 optical links.

Equalization	Distortion	Rate (Gbps)	Reach (km)	Wavelength (nm)	FEC	Tx	Rx	Ref.
FFE/DFE	bandwidth limitation & CD	56	60	1295.13	3.8×10^{-3}	EML	APD + TIA & PD + TIA	[26]
FFE/DFE	bandwidth limitation & CD	106	40	1309.49	2×10^{-4}	EML	APD	[27]
FFE/DFE	bandwidth limitation & CD	112	40	1314	1×10^{-3}	EML	PD + TIA	[28]
THP	bandwidth limitation & power fading	100	40	1300	4×10^{-3} & 2×10^{-4}	EML	PD + TIA	[72]
DD-FTN	enhanced noise	140	20	1296.2	3.8×10^{-3}	EML	PD + TIA	[35]
ID-FFE/ ID-DFE	chirp of modulator	56	43	N/A	3.8×10^{-3}	DML	PD	[37]
CR-FFE	clock offset with CD	50	40	1549.39	2×10^{-2}	DML	OA + PD	[39]
VNLE	nonlinearity	56	100	1549.81	3.8×10^{-3}	DML	OA + PD	[92]
VNLE	nonlinearity	56	70	1311.89	3.8×10^{-3}	DML	OA + PD	[93]
SVM	nonlinearity of VSCEL	60	0.05	850	3.8×10^{-3}	VSCEL	PD	[40]
SVM	level-dependent skew	50	20	1551	1×10^{-3} & 2×10^{-4}	DML	PD	[41]
ANN	nonlinearity of DML	20	18	1310	3.8×10^{-3}	DML	PD	[43]
ANN	nonlinearity	50	80	1551.35	3.8×10^{-3}	DML	OA + PD	[44]
CNN	nonlinearity	56	25	N/A	3.8×10^{-3}	DML	APD	[45]
CNN	nonlinearity	112	40	1293	2×10^{-4}	EML	PD + TIA	[46]
DNN	nonlinearity	64	4	850	2×10^{-4}	VSCEL	PD + TIA	[47]
DNN	nonlinearity	50	80	1551.35	1×10^{-3}	DML	OA + PD	[48]

At present, the standard to evaluate the equalization technologies is considering both the performance improvement and the computational complexity. The high complexity will occupy massive resources for computation, thus increasing the computational cost. The score to judge the equalization technologies can be described as:

$$\text{score} = \frac{a * \text{performance improvement}}{b * \text{computational cost}} \tag{7}$$

where a and b are the influence factors of performance improvement and computational cost, respectively. The comparison of the computational complexity and performance improvement for different equalization technologies is summarized in Table 4. Note that the main limited factor for the short-reach PAM4 system is the cost, and the goal for equalization technologies is to achieve better performance at the same computation cost. As is shown in Table 4, the SVM an NN have a higher complexity than conventional equalization, although the performance can be significantly improved. However, the computation cost is getting lower as the continuous development of the integrated circuit. So, from the long-term point of view, the equalization based on machine learning algorithms will play an increasingly important role for short-reach PAM4 optical links.

Table 4. Comparison of the computational complexity and performance improvement for different equalization technologies.

Equalization Technologies	FFE/DFE	DD-FTN	ID-FFE/ ID-DFE	CR-FFE	VNLE	Sparse VNLE	SVM	NN
Computational Complexity	low	fair	low	low	high	fair	high	high
Performance Improvement	low	fair	fair	fair	high	high	high	high

5. Conclusions and Perspective

In this paper, we have reviewed various equalization technologies for short-reach PAM4 optical links. A typical system configuration is presented and the comparisons among different transmitters and receivers are introduced. Different distortions including linear impairments, device nonlinearity and power fading effect are described, which are induced by low-cost components and need to be mitigated by pre- and post-equalization. The conventional equalizers, including FFE and DFE, are illustrated to eliminate the linear impairments such as bandwidth limitation and chromatic dispersion, while the nonlinear distortions are compensated for by conventional VNLE and sparse VNLE. The machine learning algorithms like SVM and NN are proposed to further mitigate severe nonlinear distortion and achieve significant performance improvement. Finally, a summary is given for different equalization technologies and a standard for the evaluation of equalizers is defined to consider both performance improvement and computational complexity. Additionally, as the cost of computation constantly decreases, equalization technologies based on machine learning may become the mainstream technology for next-generation short-reach PAM4 transmission.

Author Contributions: This paper was mainly wrote by H.Z. and Y.L. (Yan Li) Y.L. (Yuyang Liu) provided the idea and J.W. supervised overall project. L.Y., C.G., W.L., J.Q., H.G., X.H., Y.Z. and J.W. contributed to the reviewing and editing of the manuscript.

Funding: This paper is partly funded by National Natural Science Foundation of China (NSFC) (61875019, 61675034, 61875020, 61571067); The Fund of State Key Laboratory of IPOC (BUPT); The Fundamental Research Funds for the Central Universities.

Acknowledgments: We would like to thank the unknown reviewers that helped improve this manuscript with their valuable feedback.

Conflicts of Interest: The authors declare no conflict of interest.

Abbreviations

Acronyms		Acronyms	
ADC	Analog-to-digital converter	APD	Avalanche photodiode
BER	Bit error rate	CAP	Carrier-less amplitude and phase modulation
CBT	Complete binary tree	CD	Chromatic dispersion
CR	Clock recovery	DAC	Digital-to-analog converter
DCI	Data center interconnects	DD	Direct detection
DFE	Decision feedback equalizer	DML	Directly modulated laser
DMT	Discrete multi-tone	DSP	Digital signal processing
EML	Electro-absorption modulated lasers	FEC	Forward error correction
FFE	Feed-forward equalizer	FTN	Faster than nyquist
HD	Hard decision	ID	Intensity directed
IM	Intensity modulation	ISI	Inter-symbol interference
KK	Kramers-Kronig	LMS	Least mean squares
MLSE	Maximum likelihood sequence estimation	ML	Machine learning
MMF	Multi-mode fiber	MZM	Mach-Zehnder modulator
NN	Neural network	OA	Optical amplifier
OPBF	Optical band-pass filter	OSNR	Optical signal-to-noise ratio
PAM4	Four-level pulse amplitude modulation	PD	Photodiodes
PRBS	Pseudorandom binary sequence	QAM	quadrature amplitude modulation
ReLU	Rectified Linear Units	RLS	Recursive least squares
SCO	Sampling clock offset	SMMF	Standard single-mode fiber
SMO	Sequential minimal optimization	SPM	Self-phase modulation
SVM	Support vector machine	TIA	Trans-impedance amplifiers
VCSEL	Vertical cavity surface-emitting laser	VNLE	Volterra nonlinear equalizer
VR	virtual reality	ZF	Zero-forcing

References

1. Chen, Q.; Bahadori, M.; Glick, M.; Rumley, S.; Bergman, K. Recent advances in optical technologies for data centers: A review. *Optica* **2018**, *11*, 1354–1370. [CrossRef]
2. Cisco Visual Networking. *The Zettabyte Era—Trends and Analysis*; Cisco White Paper: San Jose, CA, USA, 2017.
3. Zhong, K.; Zhou, X.; Wang, Y.; Gui, T.; Yang, Y.; Yuan, J.; Wang, L.; Chen, W.; Zhang, H.; Man, J.; et al. Recent Advances in Short Reach Systems. In Proceedings of the Optical Fiber Communication Conference (OFC), Los Angeles, CA, USA, 19–23 March 2017; p. Tu2D.7.
4. Plant, D.V.; Morsy-Osman, M.; Chagnon, M. Optical communication systems for datacenter networks. In Proceedings of the Optical Fiber Communications Conference (OFC), Los Angeles, CA, USA, 19–23 March 2017; pp. 1–47.
5. Cartledge, J.C.; Karar, A.S. 100 Gb/s Intensity Modulation and Direct Detection. *J. Lightw. Technol.* **2014**, *16*, 2809–2814. [CrossRef]
6. Wei, J.; Cheng, Q.; Penty, R.V.; White, I.H.; Cunningham, D.G. 400 Gigabit Ethernet using advanced modulation formats: Performance, complexity, and power dissipation. *Commun. Mag.* **2015**, *2*, 182–189. [CrossRef]
7. Zhong, K.; Zhou, X.; Huo, J.; Yu, C.; Lu, C.; Lau, A.P.T. Digital signal processing for short-reach optical communications: A review of current technologies and future trends. *J. Lightw. Technol.* **2018**, *2*, 377–400. [CrossRef]
8. Xu, X.; Zhou, E.; Liu, G.N.; Zuo, T.; Zhong, Q.; Zhang, L.; Bao, Y.; Zhang, X.; Li, J.; Li, Z. Advanced modulation formats for 400-Gbps short-reach optical inter-connection. *Opt. Express* **2015**, *1*, 492–500. [CrossRef] [PubMed]
9. Lyubomirsky, I.; Ling, W.A. Advanced Modulation for Datacenter Interconnect. In Proceedings of the Optical Fiber Communications Conference (OFC), Anaheim, CA, USA, 20–24 March 2016; pp. 1–3.
10. Eiselt, N.; Griesser, H.; Wei, J.; Hohenleitner, R.; Dochhan, A.; Ortsiefer, M.; Eiselt, M.H.; Neumeyr, C.; José Vegas Olmos, J.; Monroy, I.T. Experimental Demonstration of 84 Gb/s PAM-4 Over up to 1.6 km SSMF Using a 20-GHz VCSEL at 1525 nm. *J. Lightw. Technol.* **2017**, *8*, 1342–1349. [CrossRef]
11. Sadot, D.; Dorman, G.; Gorshtein, A.; Sonkin, E.; Vidal, O. Single channel 112Gbit/sec PAM4 at 56Gbaud with digital signal processing for data centers applications. *Opt. Express* **2015**, *2*, 991–997. [CrossRef] [PubMed]
12. Tao, L.; Wang, Y.; Xiao, J.; Chi, N. Enhanced performance of 400 Gb/s DML-based CAP systems using optical filtering technique for short reach communication. *Opt. Express* **2014**, *24*, 29331–29339. [CrossRef]
13. Shi, J.; Zhang, J.; Li, X.; Chi, N.; Chang, G.; Yu, J. 112 Gb/s/λ CAP Signals Transmission over 480 km in IM-DD System. In Proceedings of the Optical Fiber Communications Conference (OFC), San Diego, CA, USA, 11–15 March 2018; pp. 1–3.
14. Zhang, L.; Zuo, T.; Mao, Y.; Zhang, Q.; Zhou, E.; Liu, G.N.; Xu, X. Beyond 100-Gb/s Transmission Over 80-km SMF Using Direct-Detection SSB-DMT at C-Band. *J. Lightw. Technol.* **2016**, *2*, 723–729. [CrossRef]
15. Sun, L.; Du, J.; Wang, C.; Li, Z.; Xu, K.; He, Z. Frequency-resolved adaptive probabilistic shaping for DMT-modulated IM-DD optical interconnects. *Opt. Express* **2019**, *9*, 12241–12254. [CrossRef]
16. Zhu, Y.; Zou, K.; Chen, Z.; Zhang, F. 224 Gb/s Optical Carrier-Assisted Nyquist 16-QAM Half-Cycle Single-Sideband Direct Detection Transmission over 160 km SSMF. *J. Lightw. Technol.* **2017**, *9*, 1557–1565. [CrossRef]
17. Chen, X.; Antonelli, C.; Chandrasekhar, S.; Raybon, G.; Mecozzi, A.; Shtaif, M.; Winzer, P. Kramers–Kronig Receivers for 100-km Datacenter Interconnects. *J. Lightw. Technol.* **2018**, *1*, 79–89. [CrossRef]
18. Mecozzi, A.; Antonelli, C.; Shtaif, M. Kramers–Kronig coherent receiver. *Optica* **2016**, *11*, 1220–1227. [CrossRef]
19. Zhong, K.; Zhou, X.; Gui, T.; Tao, L.; Gao, Y.; Chen, W.; Man, J.; Zeng, L.; Lau, A.P.T.; Lu, C. Experimental study of PAM-4, CAP-16, and DMT for 100 Gb/s short reach optical transmission systems. *Opt. Express* **2015**, *2*, 1176–1189. [CrossRef] [PubMed]
20. Shi, J.; Zhang, J.; Zhou, Y.; Wang, Y.; Chi, N.; Yu, J. Transmission Performance Comparison for 100-Gb/s PAM-4, CAP-16, and DFT-S OFDM With Direct Detection. *J. Lightw. Technol.* **2017**, *23*, 5127–5133. [CrossRef]
21. Yekani, A.; Rusch, L.A. Interplay of Bit Rate, Linewidth, Bandwidth, and Reach on Optical DMT and PAM with IMDD. *Trans. Commun.* **2019**, *4*, 2908–2913. [CrossRef]

22. Houtsma, V.; Veen, D. Optical Strategies for Economical Next Generation 50 and 100G PON. In Proceedings of the Optical Fiber Communications Conference (OFC), San Diego, CA, USA, 3–7 March 2019; p. M2B.1.
23. Nesset, D. PON roadmap. *J. Opt. Commun. Netw.* **2017**, *1*, A71–A76. [CrossRef]
24. Zhang, K.; Zhuge, Q.; Xin, H.; Xing, Z.; Xiang, M.; Fan, S.; Yi, L.; Hu, W.; Plant, D.V. Demonstration of 50Gb/s/λ Symmetric PAM4 TDM-PON with 10G-class Optics and DSP-free ONUs in the O-band. In Proceedings of the Optical Fiber Communications Conference (OFC), San Diego, CA, USA, 11–15 March 2018; pp. 1–3.
25. Ethernet Task Force. IEEE Standard P802.3bs 200 Gb/s and 400 Gb/s. Available online: http://www.ieee802.org/3/bs/index.html (accessed on 6 December 2017).
26. Zhong, K.; Zhou, X.; Wang, Y.; Huo, J.; Zhang, H.; Zeng, L.; Yu, C.; Lau, A.P.T.; Lu, C. Amplifier-Less Transmission of 56Gbit/s PAM4 over 60km Using 25Gbps EML and APD. In Proceedings of the Optical Fiber Communications Conference (OFC), Los Angeles, CA, USA, 19–23 March 2017; p. Tu2D.1.
27. Nada, M.; Yoshimatsu, T.; Muramoto, Y.; Ohno, T.; Nakajima, F.; Matsuzaki, H. 106-Gbit/s PAM4 40-km transmission using an avalanche photodiode with 42-GHz bandwidth. In Proceedings of the Optical Fiber Communications Conference (OFC), San Diego, CA, USA, 11–15 March 2018; pp. 1–3.
28. Chan, T.K.; Way, W.I. 112 Gb/s PAM4 transmission over 40km SSMF using 1. In 3 μm gain-clamped semiconductor optical amplifier. In Proceedings of the Optical Fiber Communications Conference (OFC), Los Angeles, CA, USA, 22–26 March 2015; p. Th3A.4.
29. Stojanovic, N.; Karinou, F.; Qiang, Z.; Prodaniuc, C. Volterra and Wiener Equalizers for Short-Reach 100G PAM-4 Applications. *J. Lightw. Technol.* **2017**, *21*, 4583–4594. [CrossRef]
30. Li, X.; Zhou, S.; Ji, H.; Luo, M.; Yang, Q.; Yi, L.; Hu, Y.; Li, C.; Fu, S.; Alphones, A.; et al. Transmission of 4 × 28-Gb/s PAM-4 over 160-km single mode fiber using 10G-class DML and photodiode. In Proceedings of the Optical Fiber Communications Conference (OFC), Anaheim, CA, USA, 20–24 March 2016; pp. 1–3.
31. Gao, Y.; Cartledge, J.C.; Yam, S.S.H.; Rezania, A.; Matsui, Y. 112 Gb/s PAM-4 using a directly modulated laser with linear pre-compensation and nonlinear post-compensation. In Proceedings of the European Conference and Exhibition on Optical Communication (ECOC), Dusseldorf, Germany, 18–22 September 2016; pp. 1–3.
32. Zhang, Q.; Stojanovic, N.; Prodaniuc, C.; Karinou, F.; Xie, C. Cost-effective single-lane 112 Gb/s solution for 2 mobile fronthaul and access applications. *Opt. Express* **2016**, *24*, 5720–5723. [CrossRef]
33. Zhong, K.; Chen, W.; Sui, Q.; Man, J.W.; Lau, A.P.T.; Lu, C.; Zeng, L. Low cost 400 GE transceiver for 2 km optical interconnect using PAM4 and direct detection. In Proceedings of the Asia Communications and Photonics Conference (ACP), Shanghai, China, 11–14 November 2014; p. ATh4D.2.
34. Zhong, K.; Chen, W.; Sui, Q.; Man, J.W.; Lau, A.P.T.; Lu, C.; Zeng, L. Experimental demonstration of 500 Gbit/s short reach transmission employing PAM4 signal and direct detection with 25 Gbps device. In Proceedings of the Optical Fiber Communications Conference (OFC), Los Angeles, CA, USA, 22–26 March 2015; p. TH3A.3.
35. Zhong, K.; Zhou, X.; Gao, Y.; Chen, W.; Man, J.; Zeng, L.; Lau, A.P.T.; Lu, C. 140-Gb/s 20-km Transmission of PAM-4 Signal at 1.3 μm for Short Reach Communications. *Photon. Technol. Lett.* **2015**, *16*, 1757–1760. [CrossRef]
36. Zhang, K.; Zhuge, Q.; Xin, H.; Morsy-Osman, M.; El-Fiky, E.; Yi, L.; Hu, W.; Plant, D.V. Intensity-directed Equalizer for Chirp Compensation Enabling DML-based 56Gb/s PAM4 C-band Delivery over 35.9 km SSMF. In Proceedings of the European Conference and Exhibition on Optical Communication (ECOC), Gothenburg, Sweden, 17–21 September 2017.
37. Zhang, K.; Zhuge, Q.; Xin, H.; Morsy-Osman, M.; El-Fiky, E.; Yi, L.; Hu, W.; Plant, D.V. Intensity directed equalizer for the mitigation of DML chirp induced distortion in dispersion-unmanaged C-band PAM transmission. *Opt. Express* **2017**, *23*, 28123–28135. [CrossRef]
38. Zhou, H.; Li, Y.; Gao, C.; Li, W.; Hong, X.; Wu, J. Clock Recovery and Adaptive Equalization for 50 Gbit/s PAM4 Transmission. In Proceedings of the Asia Communications and Photonics Conference (ACP), Hangzhou, China, 26–29 October 2018.
39. Zhou, H.; Li, Y.; Lu, D.; Yue, L.; Gao, C.; Liu, Y.; Hao, R.; Zhao, Z.; Li, W.; Qiu, J.; et al. Joint clock recovery and feed-forward equalization for PAM4 transmission. *Opt. Express* **2019**, *8*, 11385–11395. [CrossRef] [PubMed]
40. Chen, G.; Du, J.; Sun, L.; Zhang, W.; Xu, K.; Chen, X.; Reed, G.T.; He, Z. Nonlinear Distortion Mitigation by Machine Learning of SVM Classification for PAM-4 andPAM-8 Modulated Optical Interconnection. *J. Lightw. Technol.* **2018**, *3*, 650–657. [CrossRef]

41. Miao, X.; Bi, M.; Yu, J.; Li, L.; Hu, W. SVM-Modified-FFE Enabled Chirp Management for 10G DML-based 50Gb/s/λ PAM4 IM-DD PON. In Proceedings of the Optical Fiber Communications Conference (OFC), San Diego, CA, USA, 3–7 March 2019; p. M2B.5.
42. Chen, G.; Du, J.; Sun, L.; Zheng, L.; Xu, K.; Tsang, H.K.; Chen, X.; Reed, G.T.; He, Z. Machine Learning Adaptive Receiver for PAM-4 Modulated Optical Interconnection Based on Silicon Microring Modulator. *J. Lightw. Technol.* **2018**, *18*, 4106–4113. [CrossRef]
43. Reza, A.G.; Rhee, J.K.K. Nonlinear Equalizer Based on Neural Networks for PAM-4 Signal Transmission Using DML. *Photonics Technol. Lett.* **2018**, *15*, 1416–1419. [CrossRef]
44. Yang, Z.; Gao, F.; Fu, S.; Li, X.; Deng, L.; He, Z.; Tang, M.; Liu, D. Radial basis function neural network enabled C-band 4 × 50 Gb/s PAM-4 transmission over 80 km SSMF. *Opt. Lett.* **2018**, *15*, 3542–3545. [CrossRef] [PubMed]
45. Li, P.; Yi, L.; Xue, L.; Hu, W. 56 Gbps IM/DD PON based on 10G-Class Optical Devices with 29 dB Loss Budget Enabled by Machine Learning. In Proceedings of the Optical Fiber Communications Conference (OFC), San Diego, CA, USA, 11–15 March 2018; pp. 1–3.
46. Chuang, C.; Liu, L.; Wei, C.; Liu, J.; Henrickson, L.; Huang, W.; Wang, C.; Chen, Y.; Chen, J. Convolutional Neural Network based Nonlinear Classifier for 112-Gbps High Speed Optical Link. In Proceedings of the Optical Fiber Communications Conference (OFC), San Diego, CA, USA, 11–15 March 2018; p. W2A.43.
47. Chuang, C.Y.; Wei, C.C.; Lin, T.C.; Chi, K.L.; Liu, L.C.; Shi, J.W.; Chen, Y.; Chen, J. Employing Deep Neural Network for High Speed 4-PAM Optical Interconnect. In Proceedings of the European Conference and Exhibition on Optical Communication (ECOC), Gothenburg, Sweden, 17–21 September 2017.
48. Luo, M.; Gao, F.; Li, X.; He, Z.; Fu, S. Transmission of 4 × 50-Gb/s PAM-4 Signal over 80-km Single Mode Fiber using Neural Network. In Proceedings of the Optical Fiber Communications Conference (OFC), San Diego, CA, USA, 11–15 March 2018; p. M2F.2.
49. Wan, Z.; Li, J.; Shu, L.; Luo, M.; Li, X.; Fu, S.; Xu, K. Nonlinear equalization based on pruned artificial neural networks for 112-Gb/s SSB-PAM4 transmission over 80-km SSMF. *Opt. Express* **2018**, *8*, 10631–10642. [CrossRef]
50. Li, P.; Yi, L.; Xue, L.; Hu, W. 100Gbps IM/DD Transmission over 25km SSMF using 20Gclass DML and PIN Enabled by Machine Learning. In Proceedings of the Optical Fiber Communications Conference (OFC), San Diego, CA, USA, 11–15 March 2018; p. W2A.46.
51. Yi, L.; Li, P.; Liao, T.; Hu, W. 100Gb/s/λ IM-DD PON using 20G-class optical devices by machine learning based equalization. In Proceedings of the European Conference and Exhibition on Optical Communication (ECOC), Rome, Italy, 23–27 September 2018.
52. Estaran, J.; Rios-Müller, R.; Mestre, M.A.; Jorge, F.; Mardoyan, H.; Konczykowska, A.; Dupuy, J.Y.; Bigo, S. Artificial Neural Networks for Linear and Non-Linear Impairment Mitigation in High-Baudrate IM/DD Systems. In Proceedings of the European Conference and Exhibition on Optical Communication (ECOC), Dusseldorf, Germany, 18–22 September 2016.
53. Houtsma, V.; Chou, E.; Veen, D. 92 and 50 Gbps TDM-PON using Neural Network Enabled Receiver Equalization Specialized for PON. In Proceedings of the Optical Fiber Communications Conference (OFC), San Diego, CA, USA, 3–7 March 2019; p. M2B.6.
54. Chang, F.; Bhoja, S. New Paradigm Shift to PAM4 Signalling at 100/400G for Cloud Data Centers: A Performance Review. In Proceedings of the European Conference and Exhibition on Optical Communication (ECOC), Gothenburg, Sweden, 17–21 September 2017.
55. Karinou, F.; Stojanovic, N.; Prodaniuc, C.; Qiang, Z.; Dippon, T. 112 Gb/s PAM-4 optical signal transmission over 100-m OM4 multimode fiber for high-capacity data-center interconnects. In Proceedings of the European Conference and Exhibition on Optical Communication (ECOC), Dusseldorf, Germany, 18–22 September 2016.
56. Matsui, Y.; Pham, T.; Ling, W.A.; Schatz, R.; Carey, G.; Daghighian, H.; Sudo, T.; Roxlo, C. 55-GHz Bandwidth Short-Cavity Distributed Reflector Laser and its Application to 112-Gb/s PAM-4. In Proceedings of the Optical Fiber Communications Conference (OFC), Anaheim, CA, USA, 20–22 March 2016; p. Th5B.4.
57. Kim, B.G.; Bae, S.H.; Kim, H.; Chung, Y.C. DSP-based CSO cancellation technique for RoF transmission system implemented by using directly modulated laser. *Opt. Express* **2017**, *11*, 12152–12160. [CrossRef]
58. Wei, C.; Cheng, H.; Huang, W. On adiabatic chirp and compensation for nonlinear distortion in DML-based OFDM transmission. *J. Lightw. Technol.* **2018**, *16*, 3502–3513. [CrossRef]

59. Zhang, K.; Zhuge, Q.; Xin, H.; Hu, W.; Plant, D.V. Performance comparison of DML, EML and MZM in dispersion-unmanaged short reach transmissions with digital signal processing. *Opt. Express* **2018**, *26*, 34288–34304. [CrossRef]
60. Zhang, Q.; Stojanovic, N.; Xie, C.; Prodaniuc, C.; Laskowski, P. Transmission of single lane 128 Gbit/s PAM-4 signals over an 80 km SSMF link, enabled by DDMZM aided dispersion pre-compensation. *Opt. Express* **2016**, *21*, 24580–24591. [CrossRef] [PubMed]
61. Kai, Y.; Nishihara, M.; Tanaka, T.; Takahara, T.; Li, L.; Tao, Z.; Liu, B.; Rasmussen, J.C.; Drenski, T. Experimental Comparison of Pulse Amplitude Modulation (PAM) and Discrete Multi-tone (DMT) for Short-Reach 400-Gbps Data Communication. In Proceedings of the European Conference and Exhibition on Optical Communication (ECOC), London, UK, 22–26 September 2013; pp. 1–3.
62. Filer, M.; Searcy, S.; Fu, Y.; Nagarajan, R.; Tibuleac, S. Demonstration and performance analysis of 4 Tb/s DWDM metro-DCI system with 100 G PAM4 QSFP28 modules. In Proceedings of the Optical Fiber Communications Conference (OFC), Los Angeles, CA, USA, 19–23 March 2017; pp. 1–3.
63. Zhang, J.; Wey, J.S.; Shi, J.; Yu, J.; Tu, Z.; Yang, B.; Yang, W.; Guo, Y.; Huang, X.; Ma, Z. Experimental Demonstration of Unequally Spaced PAM-4 Signal to Improve Receiver Sensitivity for 50-Gbps PON with Power-Dependent Noise Distribution. In Proceedings of the Optical Fiber Communications Conference (OFC), San Diego, CA, USA, 11–15 March 2018; pp. 1–3.
64. Castro, J.M.; Pimpinella, R.J.; Kose, B.; Huang, Y.; Novick, A.; Lane, B. Eye Skew Modeling, Measurements and Mitigation Methods for VCSEL PAM-4 Channels at Data Rates over 66 Gb/s. In Proceedings of the Optical Fiber Communications Conference (OFC), Los Angeles, CA, USA, 19–23 March 2017; p. W3G.3.
65. Rath, R.; Clausen, D.; Ohlendorf, S.; Pachnicke, S.; Rosenkranz, W. Tomlinson–Harashima Precoding for Dispersion Uncompensated PAM-4 Transmission with Direct-Detection. *J. Lightw. Technol.* **2017**, *18*, 3909–3917. [CrossRef]
66. Tang, X.; Liu, S.; Xu, X.; Qi, J.; Guo, M.; Zhou, J.; Qiao, Y. 50-Gb/s PAM4 over 50-km Single Mode Fiber Transmission Using Efficient Equalization Technique. In Proceedings of the Optical Fiber Communications Conference (OFC), San Diego, CA, USA, 3–7 March 2019; p. W2A.45.
67. Proakis, J.G.; Manolakis, D.G. *Digital Signal. Processing: Principles, Algorithms and Applications*, 4th ed.; Pearson Education: London, UK, 2007.
68. Haykin, S.S. *Adaptive Filter Theory*, 5th ed.; Pearson Education: Hamilton, ON, Canada, 2005.
69. Proakis, J.G. *Digital Communications*, 4th ed.; The McGraw-Hill Companies: San Diego, CA, USA, 2008.
70. Rath, R.; Rosenkranz, W. Tomlinson-Harashima Precoding for Fiber-Optic Communication Systems. In Proceedings of the European Conference and Exhibition on Optical Communication (ECOC), London, UK, 22–26 September 2013.
71. Matsumoto, K.; Yoshida, Y.; Maruta, A.; Kanno, A.; Yamamoto, N.; Kitayama, K. On the impact of Tomlinson-Harashima precoding in optical PAM transmissions for intra-DCN communication. In Proceedings of the Optical Fiber Communications Conference (OFC), Los Angeles, CA, USA, 19–23 March 2017; p. Th3D.7.
72. Kikuchi, N.; Hirai, R.; Fukui, T. Application of Tomlinson-Harashima Precoding (THP) for Short-Reach Band-Limited Nyquist PAM and Faster-Than-Nyquist PAM Signaling. In Proceedings of the Optical Fiber Communications Conference (OFC), San Diego, CA, USA, 11–15 March 2018; pp. 1–3.
73. Kikuchi, N.; Hirai, R.; Fukui, T.; Takashima, S. Modulator Non-Linearity Compensation in Tomlinson-Harashima Precoding (THP) for Short-Reach Nyquist- and Faster-than-Nyquist (FTN) IM/DD PAM Signaling. In Proceedings of the European Conference and Exhibition on Optical Communication (ECOC), Rome, Italy, 23–27 September 2018.
74. Zhou, X.; Chen, X.; Zhou, W.; Fan, Y.; Zhu, H.; Li, Z. All-Digital Timing Recovery and Adaptive Equalization for 112 Gbit/s POLMUX-NRZ-DQPSK Optical Coherent Receivers. *J. Opt. Commun. Netw.* **2010**, *11*, 984–990. [CrossRef]
75. Mathews, V.J. Adaptive polynomial filters. *IEEE Signal Process. Mag.* **1991**, *3*, 10–26. [CrossRef]
76. Zhu, A.; Pedro, J.C.; Cunha, T.R. Pruning the Volterra series for behavioral modeling of power amplifiers using physical knowledge. *Trans. Microw. Theory Tech.* **2007**, *5*, 813–821. [CrossRef]
77. Li, X.; Zhou, S.; Gao, F.; Luo, M.; Yang, Q.; Mo, Q.; Yu, Y.; Fu, S. 4 × 28 Gb/s PAM4 long-reach PON using low complexity nonlinear compensation. In Proceedings of the Optical Fiber Communications Conference (OFC), Los Angeles, CA, USA, 19–23 March 2017; p. M3H.4.

78. Lu, S.; Wei, C.; Chuang, C.; Chen, Y.; Chen, J. 81.7% Complexity Reduction of Volterra Nonlinear Equalizer by Adopting L1 Regularization Penalty in an OFDM Long-Reach PON. In Proceedings of the European Conference and Exhibition on Optical Communication (ECOC), Gothenburg, Sweden, 17–21 September 2017.
79. Huang, W.; Chang, W.; Wei, C.; Liu, J.; Chen, Y.; Chi, K.; Wang, C.; Shi, J.; Chen, J. 93% Complexity Reduction of Volterra Nonlinear Equalizer by L1-Regularization for 112-Gbps PAM-4 850-nm VCSEL Optical Interconnect. In Proceedings of the Optical Fiber Communications Conference (OFC), San Diego, CA, USA, 11–15 March 2018; pp. 1–3.
80. Ge, L.; Zhang, W.; Liang, C.; He, Z. Threshold based Pruned Retraining Volterra Equalization for PAM-4 100 Gbps VCSEL and MMF Based Optical Interconnects. In Proceedings of the Asia Communications and Photonics Conference (ACP), Hangzhou, China, 26–29 October 2018.
81. Samuel, A.L. Some Studies in Machine Learning Using the Game of Checkers. *J. Res. Dev.* **1959**, *3*, 210–229. [CrossRef]
82. Zibar, D.; Piels, M.; Jones, R.; Schäeffer, C.G. Machine learning techniques in optical communication. *J. Lightw. Technol.* **2016**, *6*, 1442–1452. [CrossRef]
83. Sebald, D.J.; Bucklew, J.A. Support vector machine techniques for nonlinear equalization. *IEEE Trans. Signal Process.* **2000**, *11*, 3217–3226. [CrossRef]
84. Chen, G.; Sun, L.; Xu, K.; Du, J.; He, Z. Machine learning of SVM classification utilizing complete binary tree structure for PAM-4/8 optical interconnection. In Proceedings of the Optical Interconnects Conference (OI), Santa Fe, NM, USA, 5–7 June 2017.
85. Ito, K.; Niwa, M.; Ueda, K.; Mori, Y.; Hasegawa, H.; Sato, K.I. Impairment mitigation in non-coherent optical transmission enabled with machine learning for intra-datacenter networks. In Proceedings of the Next-Generation Optical Networks for Data Centers and Short-Reach Links, San Francisco, CA, USA, 28 January 2017.
86. Lyubomirsky, I. Machine learning equalization techniques for high speed PAM4 fiber optic communication systems. In *CS229 Final Project Report*; Stanford University: Stanford, CA, USA, 2015.
87. Zeng, Z.; Yu, H.; Xu, H.; Xie, Y.; Gao, J. Fast training support vector machines using parallel sequential minimal optimization. In Proceedings of the International Conference on Intelligent System and Knowledge Engineering, Xiamen, China, 17–19 November 2008; pp. 997–1001.
88. Specht, D.F. A general regression neural network. *Trans. Neural Netw.* **1991**, *6*, 568–576. [CrossRef]
89. Eriksson, T.A.; Bülow, H.; Leven, A. Applying Neural Networks in Optical Communication Systems: Possible Pitfalls. *Photonics Technol. Lett.* **2017**, *23*, 2091–2094. [CrossRef]
90. Wang, D.; Zhang, M.; Li, J.; Li, Z.; Li, J.; Song, C.; Chen, X. Intelligent constellation diagram analyzer using convolutional neural network-based deep learning. *Opt. Express* **2017**, *15*, 17150–17166. [CrossRef] [PubMed]
91. Khan, F.N.; Zhou, Y.; Lau, A.P.T.; Lu, C. Modulation format identification in heterogeneous fiber-optic networks using artificial neural networks. *Opt. Express* **2012**, *11*, 12422–12431. [CrossRef] [PubMed]
92. Zhou, S.; Li, X.; Yi, L.; Yang, Q.; Fu, S. Transmission of 2 × 56 Gb/s PAM-4 signal over 100 km SSMF using 18 GHz DMLs. *Opt. Lett.* **2016**, *8*, 1805–1808. [CrossRef] [PubMed]
93. Gao, F.; Zhou, S.; Li, X.; Fu, S.; Deng, L.; Tang, M.; Liu, D.; Yang, Q. 2 × 64 Gb/s PAM-4 transmission over 70 km SSMF using O-band 18G-class directly modulated lasers (DMLs). *Opt. Express* **2017**, *7*, 7230–7237. [CrossRef] [PubMed]

Article

Experimental Investigation of 400 Gb/s Data Center Interconnect Using Unamplified High-Baud-Rate and High-Order QAM Single-Carrier Signal

Yang Yue *, **Qiang Wang** and **Jon Anderson**

Juniper Networks, 1133 Innovation Way, Sunnyvale, CA 94089, USA; qiwang.thresh@gmail.com (Q.W.); jonanderson@juniper.net (J.A.)
* Correspondence: yyue@juniper.net; Tel.: +1-408-745-2000

Received: 16 March 2019; Accepted: 13 June 2019; Published: 15 June 2019

Abstract: In this article, we review the latest progress on data center interconnect (DCI). We then discuss different perspectives on the 400G pluggable module, including form factor, architecture, digital signal processing (DSP), and module power consumption, following 400G pluggable optics in DCI applications. Next, we experimentally investigate the capacity-reach matrix for high-baud-rate and high-order quadrature amplitude modulation (QAM) single-carrier signals in the unamplified single-mode optical fiber (SMF) link. We show that the 64 GBd 16-QAM, and 64-QAM signals can potentially enable 400 Gb/s and 600 Gb/s DCI application for 40 km and beyond of unamplified fiber link.

Keywords: 400 Gigabit Ethernet; coherent communications; data center interconnect; fiber optics links and subsystems; optical communications; QSFP-DD transceiver

---

1. Introduction

To meet the ever-growing demands on the internet bandwidth, the Ethernet speed has evolved at an astonishing pace, from 10 Megabit Ethernet to 100 Gigabit Ethernet (100 GbE) within the past 30 years [1]. In 2017, 400GbE standard was ratified by the IEEE P802.3bs Task Force [2]. This built the foundation for industrial deployment of 400GbE in the global network. It is anticipated that 400GbE will be rapidly adopted over the next few years [3]. Recently, 400GbE pluggable optics is under development to satisfy the demand on bandwidth-consuming applications, including data-center-based cloud services [4–6].

For intra data center interconnect (DCI) application, optical direct detection with four-level pulse amplitude modulation (PAM-4) has been widely adopted to enable the low cost 400GbE optical transceivers, which can carry up to 400 Gb/s Ethernet data over either parallel fibers or multiple wavelengths. Direct detection PAM-4 optics has also been standardized by Multi-Source Agreement (MSA) groups to achieve link distance from 500 m to 10 km (e.g., 400G-FR8, 400G-DR4, 400G-LR4, etc.) [7]. For inter DCI application, pluggable 400G digital coherent optics (DCO) with single-carrier dual-polarization quadrature amplitude modulation (DP-QAM) signal is being proposed to achieve 40 km and beyond fiber link [8,9]. PAM-4 signal with direct detection only uses the intensity of the light to carry information. Instead, DP-QAM signal with coherent detection technology can utilize a few degrees of freedom for photons, including polarization, intensity, and phase. Consequently, coherent technology can transmit a single-carrier 400 Gb/s signal with significantly improved spectral efficiency. Furthermore, its receiver sensitivity can be significantly improved over the one of intensity modulation with direct detection (IM/DD) systems.

In this article, we first introduce the application scenarios of pluggable optics in DCI applications. Next, we discuss different perspectives on the 400G pluggable module, including form factor, architecture, digital signal processing (DSP), and module power consumption. Then, we experimentally investigate the fiber transmission performance of 16-QAM and 64-QAM single-carrier signals at various baud rates, aiming at 400 Gb/s DCI applications with unamplified link. Here, we consider 25% overhead (OH) soft-decision forward error correction (SD-FEC), which gives a 4.2×10^{-2} bit error ratio (BER) threshold [10]. For 64 GBd 16-QAM signals with 12 dB and 18 dB link margins, it is possible to transmit 400 Gb/s net data rate over 40 km and 80 km unamplified fiber link, respectively. With 18 dB link margin, 64 GBd 64-QAM signal can produce greater than 600 Gb/s net data rate for an unamplified link as long as 40 km. We further study the impact of transmitter (Tx) side power on the transmission performance and obtain a high-baud-rate QAM signal transmission matrix in unamplified link. For various signals, extra 6-dB more power can extend the fiber transmission reach to approximately 30 km. It is thus critically important to increase the Tx output power or improve the Rx sensitivity, under the available power and space constrains of the DCO pluggable module.

2. 400G Data Center Interconnect Use Cases

Data centers have been proliferated over the past few years, which are driven by more and more bandwidth-hungry applications, including enterprise digital transformation, cloud services, and the coming 5G applications. In such an era, enterprises, service providers, and cloud providers need a DCI network that can quickly adapt to the changes in business and operational demands. To meet the above requirements, data center networks are expanding in number and capacity, driving the need for higher-speed DCI solutions. Top cloud service providers, such as Amazon, Facebook, Google, and Microsoft, have already adopted 100 Gb/s and will quickly transfer to 400 Gb/s and beyond.

Figure 1 illustrates typical application scenarios of 400 Gb/s pluggable optics for DCI. As shown in Figure 1a, the reach of an unamplified link is power limited, and it mainly depends on the optical transmitter output power, the fiber link loss, and the optical receiver sensitivity. For the amplified dense wavelength division multiplexing (DWDM) link displayed in Figure 1b, it is noise limited, and its reach depends on the available optical signal to noise ratio (OSNR) at the optical receiver.

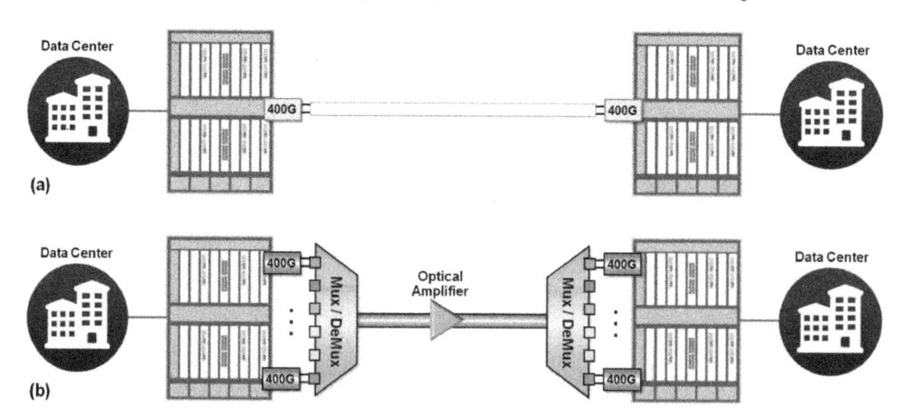

Figure 1. Typical application scenarios of 400 Gb/s pluggable optics for data center interconnect. (**a**) Unamplified point-to-point applications for 40 km reach; (**b**) amplified dense wavelength division multiplexing (DWDM) applications for 80–120 km reach.

3. 400G Pluggable Module Form Factors and Power Requirement

Due to the emerging bandwidth-hungry applications, the photonics communications industry is unceasingly driven towards higher-capacity, smaller-size, and lower-power solutions. Standardization organizations, including the Institute of Electrical and Electronics Engineers (IEEE) and the Optical Internetworking Forum (OIF), are leading the industrial transformation of photonics communications society from using discrete optical components to utilizing photonics integration enabled packet optics [11,12]. Lately, line and client-side industrial trials on interoperability have been demonstrated successfully using 100 Gb/s CFP2 and 400 Gb/s CFP8 pluggable optical modules, respectively [13–17]. Nowadays, the industry is marching towards 400 Gb/s pluggable optical modules with even more compact form factors. The first debuts of 400 Gb/s client and line-side pluggable QSFP modules are planned for year 2019 and 2020.

Based on the DSP's physical location, there are two types of pluggable coherent modules: analog coherent optics (ACO) and DCO. For 400G pluggable coherent transceivers, the industry has reached an agreement on implementing DCO, which has the DSP inside the module. A few form factors have emerged recently as the candidates for 400G DCO pluggable modules, such as CFP2, CFP8, OSFP, and QSFP-DD [18–22], whose mechanical dimensions and power consumptions are shown in Figure 2. These transceivers are equipped with eight pairs of differential signal pins, which can support 400G transmission speed using 50G PAM-4 electrical interface [23]. One can see that CFP2 and CFP8 have larger physical dimensions than those of OSFP and QSFP-DD, thus, they have more room for additional components and allow a larger power budget. Consequently, OSFP and QSFP-DD are targeted for the DCI application, while CFP2 and CFP8 are planned for the long-haul applications. The line-card fitting into one rack unit (RU) space can accommodate 36 QSFP-DD pluggable modules. The total capacity of the line-card is 14.4 Tb/s. Currently, the industry is converging behind QSFP-DD due to its high port density in the front panel, its backward compatibility with QSFP28/QSFP+, and thus a larger ecosystem.

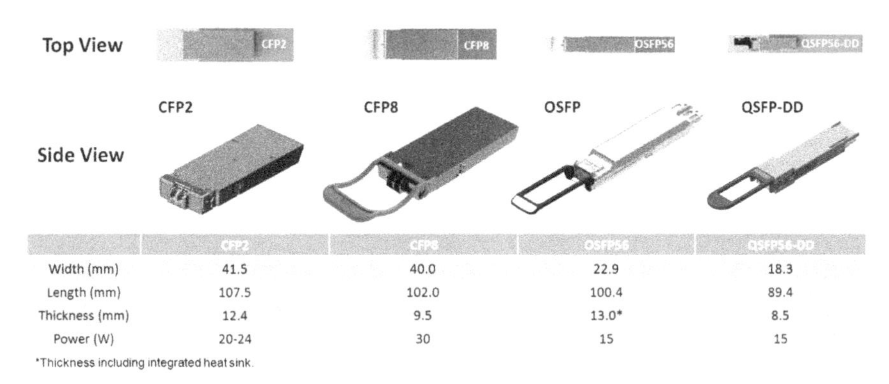

Figure 2. 400G pluggable coherent transceiver form factors. CFP: C form-factor, OSFP: octal small form factor, QSFP-DD: quad small form factor double density.

Figure 3 illustrates the typical architecture of a 400G pluggable coherent transceiver. There are three functional blocks inside the module, one DSP chip, one transmitter and receiver optical subassembly (TROSA), and one power/control unit. Besides the power and control signal, the host side also provides up to 400GbE client traffic to the pluggable coherent module through eight pairs of differential signal pins. 400 Gb/s single-carrier signal (>60 GBd 16-QAM) can thus be generated out of the Tx output. The other port of receiver (Rx) input captures the received signal, and coherent detection is performed by the TROSA and DSP.

The pluggable coherent optics starts from 100 Gb/s CFP. During the past few years, its evolution of density has been quite amazing. Meanwhile, a lot of challenges to the whole industry have been triggered. Power consumption is always one of the key constrains for industrial applications. As shown in Figure 2, the requirement on power consumption for 400 Gb/s QSFP-DD DCO is less than 15 W. Over the years, the power consumption of the coherent DSP has reduced significantly [24–26]. Table 1 illustrates the evolution of the power consumption of coherent DSP. One can see that, with 40 nm complementary metal-oxide-semiconductor (CMOS) process, 100G coherent traffic consumes approximately 50 W power. Nowadays, with the latest 7 nm CMOS technique, the industry can bring the power consumption of DSP down to less than 2 W, which is larger than a 20 times reduction. This tremendous improvement can potentially fit the 400G-ZR QSFP-DD DCO under a 15 W power envelop. As shown in Table 2, DSP, TROSA, and the other components are allocated only 7 W, 6–7 W, and 1–2 W power budget, respectively. In such tight power and size constrains, the industry has adopted 10–12 dB link budget in the power limited unamplified case for up to 40 km reach.

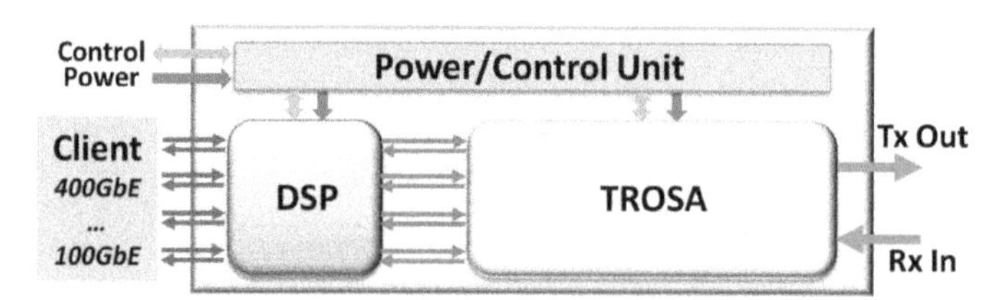

Figure 3. Typical architecture of 400G pluggable coherent transceiver. DSP: digital signal processing, TROSA: transmitter and receiver optical subassembly, Tx: transmitter, Rx: receiver.

Table 1. Coherent digital signal processing (DSP) power consumption evolution. CMOS: complementary metal-oxide-semiconductor.

DSP	2012	2015	2017	2019
CMOS node	40 nm	28 nm	16 nm	7 nm
Power/100G	<50 W	<20 W	<7 W	<2 W

Table 2. 400G-ZR QSFP-DD power consumption. TROSA: transmitter and receiver optical subassembly, QSFP-DD: quad small form factor double density.

Module Power	DSP	TROSA	Other
QSFP-DD	7 W	6–7 W	1–2 W

4. Experimental Setup

Figure 4 shows our experimental setup to investigate single-carrier high-baud-rate and high-order QAM signal transmission in unamplified fiber link. A 92 GS/s digital-to-analog converter (DAC) with 32 GHz analog bandwidth is used to generate electrical signals for the four tributaries of dual-polarization (DP) in-phase and quadrature (IQ) modulator. The trace loss between the DAC and the DP-IQ modulator is compensated in the digital domain by a finite impulse response (FIR) filter [27]. The root-raise cosine filter with a roll-off factor of 0.2 was applied in the following experiment. We used two tunable narrow-linewidth (<100 kHz) lasers with 16 dBm maximal output power, and set them to 1550.12 nm (i.e., 193.4 THz). One is with maximal power for the optical input to the DP-IQ modulator, and the other is set to 12 dBm to act as the local oscillator (LO) for the coherent receiver. The laser to the modulator is separated into two orthogonal polarizations, and each polarization is modulated by the corresponding IQ signals. An automatic bias control (ABC) is used to control the bias voltage applied to the DP-IQ modulator. At the Rx side, the optical signal is first de-modulated by using a 90° heterodyne coherent detector. Four linear trans-impedance amplifiers (TIAs) with differential outputs are set to automatic gain control (AGC) mode. This minimizes the BER performance dependence on the LO power, and four tributaries of analog electrical signals are generated. Then, an 8-bit analog-to-digital converter (ADC) with 63 GHz analog bandwidth and 4.5 effective number of bits (ENOB) samples the analog signals into discrete digital samples at 160 GS/s for post signal processing. In the DSP, chromatic dispersion (CD) is first compensated. Then polarization mode dispersion (PMD) is compensated and polarization tributaries are de-multiplexed by a blind equalization using the radius directed equalization (RDE). A 21-tap equalizer is implemented in the time domain with a Ts/2 spacing [28].

Furthermore, the blind estimation algorithm is used to estimate and compensate the carrier frequency offset between the Tx signal and the LO laser, followed by the carrier phase noise [29].

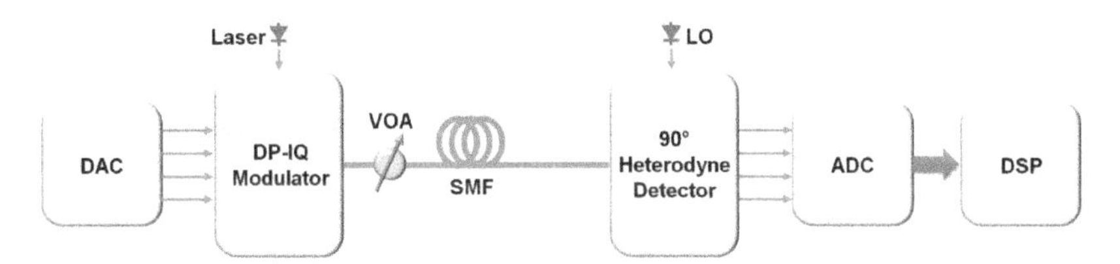

Figure 4. Experimental setup. DAC: digital-to-analog converter, IQ: in-phase (I) and quadrature (Q), VOA: variable optical attenuator, SMF: single-mode fiber, LO: local oscillator, ADC: analog-to-digital converter, DSP: digital signal processing.

In our experiment, we tried to emulate the power limited unamplified case, which is being adopted by the industry with 10–12 dB link margin. According to the receiver sensitivity of 64 GBd 16-QAM signal, we adjusted the variable optical attenuator (VOA) to set the transmitter optical output power to a reference point $P_{Tx\ Ref}$, which is 12 dB greater than the receiver sensitivity at a BER of 4.2×10^{-2}. In our experiment, we generated 16-QAM and 64-QAM with symbol rates from 45 to 86 GBd. By adjusting the VOA, we also increased the Tx optical power into the single-mode optical fiber (SMF) by 3 dB and 6 dB. Optical 16-QAM and 64-QAM signals at various baud rates were then transmitted over an SMF link with distance from 10 to 110 km. Different combinations of a few fiber sections were used to achieve specific fiber length, and their corresponding losses are illustrated in Table 3.

Table 3. Fiber loss characterization.

Length (km)	20	40	60	80	90	100	110
Loss (dB)	4.7	7.8	12.6	15.5	17.6	20.0	22.8

5. Fiber Transmission Performance Results and Discussion

In this section, we first study the fiber transmission performance with high-baud-rate 16-QAM signals in the unamplified link. Then, we extend our experiment to the high-baud-rate 64-QAM signals. Finally, we summarize a transmission matrix showing capacity and distance for 16-QAM and 64-QAM signals at various baud rates.

5.1. High-Baud-Rate 16-QAM

For 400 Gb/s DCI applications, the industry is adopting DP-16QAM signals. With additional FEC overhead, the DP-16QAM signal can also fuel the 400 Gb/s metro, regional, and even long-haul DWDM applications. Figure 5a displays the fiber transmission performance of the 56 GBd 16-QAM signal with different Tx side optical power. As the optical link is power-limited, the transmission performance can be improved for longer transmission by increasing the output power of Tx. With 6-dB more optical power than $P_{Tx\ Ref}$, the 56 GBd 16-QAM signal can transmit over 100 km with a pre-FEC BER of 2.5×10^{-2}. From the figure, one can see that, with 3-dB more optical power than $P_{Tx\ Ref}$, it is sufficient to achieve 80 km error-free transmission after FEC. With different optical power of the Tx side, the fiber transmission performance of the 16-QAM signals at 64 GBd is shown in Figure 5b. Using 25% OH SD-FEC, the 64 GBd 16-QAM signal can still carry over 400 Gb/s net data rate. Experimental result shows that a larger than 50 km unamplified optical communication link can be achieved using $P_{Tx\ Ref}$. With 6-dB more optical power than $P_{Tx\ Ref}$, the system can achieve larger than 90 km transmission in the unamplified link. Consequently, enhanced output power of Tx or improved sensitivity of Rx would be beneficial for 80 km extended long reach (ZR) application [9].

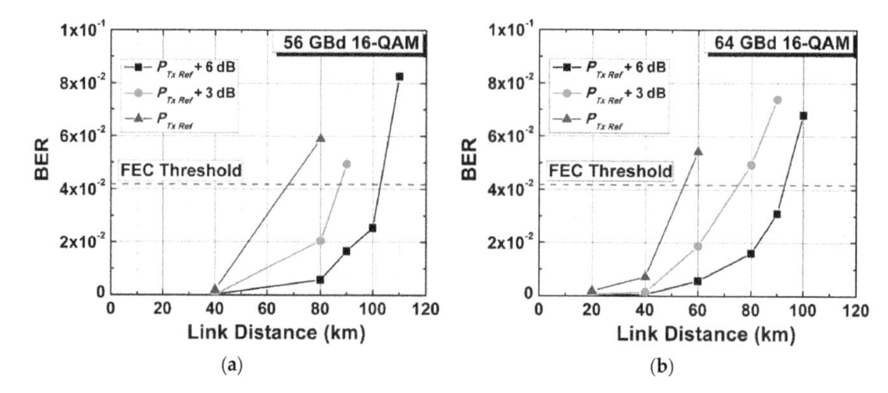

(a) (b)

Figure 5. Bit error ratio (BER) as a function of transmission distance in unamplified point-to-point link for (**a**) 56 GBd 16-QAM signals; (**b**) 64 GBd 16-QAM signals.

Furthermore, we compare 16-QAM signal transmission performance at various baud rates with $P_{Tx\ Ref}$ and 6-dB more Tx side optical power, respectively. As shown in Figure 6a,b, we transmitted the 16-QAM signal with up to 86 GBd symbol rate. For the 86 GBd 16-QAM signals with $P_{Tx\ Ref}$ and 6-dB more Tx side optical power, their reaches are around 40 km and 75 km, which are mainly limited by the increased receiver sensitivity (i.e., more receiving optical power is required) for higher baud rate system.

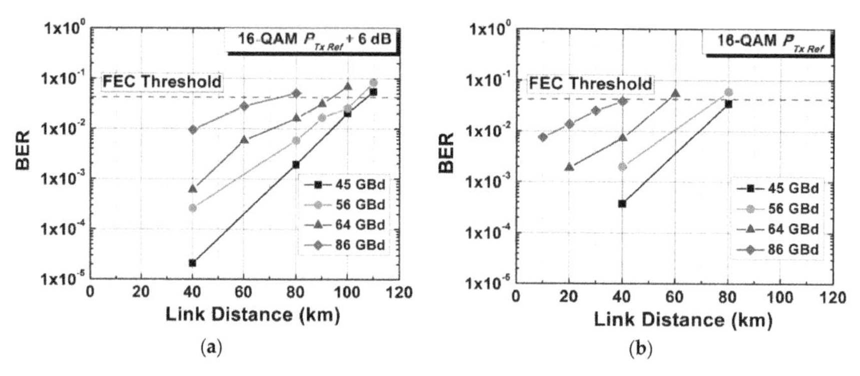

(a) (b)

Figure 6. BER as a function of transmission distance in unamplified point-to-point link for 16-QAM signals with (**a**) 6-dB more transmitter output power than the reference point; (**b**) transmitter output power at the reference point.

5.2. High-Baud-Rate 64-QAM

Compared with 16-QAM, the 64-QAM signal has 1.5× more spectral efficiency, and thus can carry more information bits. Figure 7a,b illustrates the fiber transmission performance of the 45 GBd and 64 GBd 64-QAM signals over different link distances. Compared with the 16-QAM, the 64-QAM signal has a smaller Euclidean distance between neighboring constellation points. It is thus much more susceptible to noise sources, such as laser phase noise and Rx electrical noise. In addition, compared with 16-QAM, the nonlinearity introduced by the transfer function of the IQ modulator has more impact on the 64-QAM signals. In our experiment, the modulation nonlinearity is mitigated by adjusting the electrical driving signal. As shown in Figure 7b, with 6-dB more optical power of the Tx side optical power than $P_{Tx\ Ref}$, the 64 GBd 64-QAM signal can transmit up to 40 km in the unamplified link, which gives greater than 600 Gb/s net data rate.

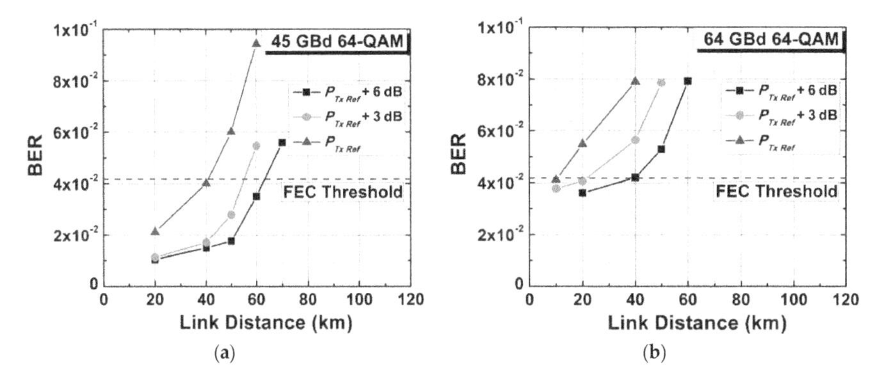

(a) (b)

Figure 7. BER as a function of transmission distance in unamplified point-to-point link for (**a**) 45 GBd 64-QAM signals; (**b**) 64 GBd 64-QAM signals.

5.3. High-Baud-Rate QAM Transmission Matrix

After transmitting 16-QAM and 64-QAM signals with various baud rates and different Tx side powers, we summarized the single-carrier signal transmission matrix, using the net data rate and achievable unamplified link distance. As shown in Figure 8, the link distance data points are interpolated at 4.2×10^{-2} SD-FEC threshold. The net data rate calculates only the information bits, by excluding the 25% FEC OH. For each combination of baud rate and QAM order, we chose two points of optical power at the Tx side, $P_{Tx\ Ref}$ and 6-dB more optical power than $P_{Tx\ Ref}$. For the 64 GBd 16-QAM signals, these conditions provide 12 dB and 18 dB link margin, respectively. From Figure 8 one can see that, for the same modulation format, lowering the baud rate gives an extended reach at the sacrifice of the net data rate. Similarly, for the same baud rate, increasing the QAM order gives a higher net data rate at the expense of the system's reach. For $P_{Tx\ Ref}$ Tx side optical power, the 45 GBd 16-QAM signal can still achieve 80 km reach in the unamplified link, which provides 288 Gb/s net data rate. For the Tx side with 6-dB more optical power than $P_{Tx\ Ref}$, the 64 GBd 16-QAM signal can achieve both 80 km transmission and larger than 400 Gb/s net data rate. In such a condition, the 64 GBd 64-QAM signal can produce greater than 600 Gb/s net data rate for an unamplified link as long as 40 km. By comparing the two conditions of optical power at the Tx side, one can see that 6-dB more power can extend the fiber transmission reach to approximately 30 km. Consequently, it is critically important to increase the output power at the Tx side or improve the sensitivity at the Rx side. Increasing the laser's power, reducing the modulation loss, using two separate lasers for Tx and LO are a few potential options.

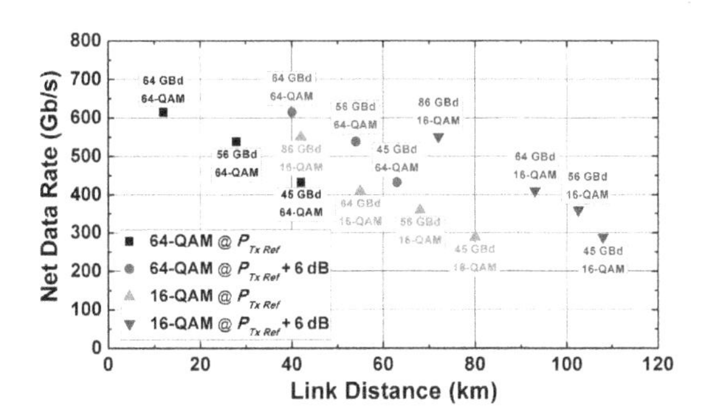

Figure 8. Net data rate as a function of transmission distance for 16-QAM and 64-QAM signals at various baud rates.

6. Conclusions

In this article, we introduce 400GbE DCI using pluggable optics. Different aspects of the 400G pluggable module, including form factor, architecture, and power consumption, are discussed. Experimental investigation shows that the 64 GBd 16-QAM and 64-QAM signals can potentially enable 400 Gb/s and 600 Gb/s DCI application for 40 km and beyond unamplified fiber link. Capacity-reach matrix is finally generated with different baud rates, QAM order, and Tx side power in the unamplified fiber link. We find that increasing the Tx output power or improving the Rx sensitivity is the key for better system performance in the power-limited coherent fiber link. A 6-dB more transmitted power can extend the fiber transmission reach to approximately 30 km for both modulation formats investigated. It is thus important for the industry to develop accordingly, under the power and space constrains of the 400G DCO pluggable module.

Author Contributions: Conceptualization, Y.Y. and Q.W.; methodology, Y.Y. and Q.W.; software, Y.Y. and Q.W.; validation, Y.Y. and Q.W.; formal analysis, Y.Y. and Q.W.; investigation, Y.Y. and Q.W.; resources, Y.Y., Q.W. and J.A.; data curation, Y.Y. and Q.W.; writing—original draft preparation, Y.Y. and Q.W.; writing—review and editing, Y.Y., Q.W. and J.A.; visualization, Y.Y. and Q.W.; supervision, J.A.

Funding: This research received no external funding.

Acknowledgments: The authors gratefully acknowledge Xuan He and Jeffery J. Maki for the fruitful discussion on the work. The authors also gratefully acknowledge vigorous encouragement and sturdy support on innovation from Domenico Di Mola at Juniper Networks.

Conflicts of Interest: The authors declare no conflicts of interest.

References

1. 40 Gb/s and 100 Gb/s Fiber Optic Task Force. Available online: http://www.ieee802.org/3/bm/ (accessed on 26 January 2019).
2. IEEE P802.3bs 400 Gb/s Ethernet Task Force. Available online: www.ieee802.org/3/bs/ (accessed on 1 March 2019).
3. Rokkas, T.; Neokosmidis, I.; Tomkos, I. Cost and Power Consumption Comparison of 400 Gbps Intra-Datacenter Transceiver Modules. In Proceedings of the 2018 International Conference on Transparent Optical Networks (ICTON), Bucharest, Romania, 1–5 July 2018.
4. Urata, R.; Liu, H.; Zhou, X.; Vahdat, A. Datacenter Interconnect and Networking: From Evolution to Holistic Revolution. In Proceedings of the 2018 Optical Fiber Communications Conference and Exposition (OFC), San Diego, CA, USA, 11–15 March 2018.
5. Baveja, P.P.; Li, M.; Wang, D.; Hsieh, C.; Zhang, H.; Ma, N.; Wang, Y.; Lii, J.; Liang, Y.; Wang, C.; et al. 56 Gb/s PAM-4 Directly Modulated Laser for 200G/400G Data-Center Optical Links. In Proceedings of the 2018 Optical Fiber Communications Conference and Exposition (OFC), San Diego, CA, USA, 11–15 March 2018.
6. Yue, Y.; Wang, Q.; Maki, J.J.; Anderson, J. Enabling Technologies in Packet Optics. In Proceedings of the 2018 Asia Communications and Photonics Conference (ACP), Hangzhou, China, 26–29 October 2018.
7. IEEE P802.3bs 200 Gb/s and 400 Gb/s Ethernet Task Force. Available online: www.ieee802.org/3/bs/ (accessed on 26 January 2019).
8. Sone, Y.; Nishizawa, H.; Yamamoto, S.; Fukutoku, M.; Yoshimatsu, T. Systems and technologies for high-speed inter-office/datacenter interface. In Proceedings of the 2017 SPIE Photonics West, San Francisco, CA, USA, 28 January–2 February 2017.
9. OIF 400ZR. Available online: http://www.oiforum.com/technical-work/hot-topics/400zr-2/ (accessed on 20 May 2019).
10. Zhu, Y.; Li, A.; Peng, W.; Kan, C.; Li, Z.; Chowdhury, S.; Cui, Y.; Bai, Y. Spectrally-Efficient Single-Carrier 400G Transmission Enabled by Probabilistic Shaping. In Proceedings of the 2017 Optical Fiber Communications Conference and Exposition (OFC), Los Angeles, CA, USA, 19–23 March 2017.
11. Isono, H. Latest standardization trends for client and networking optical transceivers and its future directions. In Proceedings of the 2018 optoelectronics, photonic materials and devices conference (SPIE OPTO), San Francisco, CA, USA, 27 January–1 February 2018.
12. Cole, C. Beyond 100G client optics. *IEEE Commun. Mag.* **2012**, *50*, s58–s66. [CrossRef]

13. Garrafa, N.; Salome, O.; Mueller, T.; Carcelen, O.P.; Calabretta, G.; Carretero, N.; Galimberti, G.; Keck, S.; López, V.; van den Borne, D. Multi-vendor 100G DP-QPSK line-side interoperability field trial over 1030 km. In Proceedings of the 2017 Optical Fiber Communications Conference and Exposition (OFC), Los Angeles, CA, USA, 19–23 March 2017.
14. Nelson, L.E.; Zhang, G.; Padi, N.; Skolnick, C.; Benson, K.; Kaylor, T.; Iwamatsu, S.; Inderst, R.; Marques, F.; Fonseca, D.; et al. SDN-Controlled 400GbE end-to-end service using a CFP8 client over a deployed, commercial flexible ROADM system. In Proceedings of the 2017 Optical Fiber Communications Conference and Exposition (OFC), Los Angeles, CA, USA, 19–23 March 2017.
15. Birk, M.; Nelson, L.E.; Zhang, G.; Cole, C.; Yu, C.; Akashi, M.; Hiramoto, K.; Fu, X.; Brooks, P.; Schubert, A.; et al. First 400GBASE-LR8 interoperability using CFP8 modules. In Proceedings of the 2017 Optical Fiber Communications Conference and Exposition (OFC), Los Angeles, CA, USA, 19–23 March 2017.
16. Nelson, L.E. Advances in 400 Gigabit Ethernet Field Trials. In Proceedings of the 2018 Optical Fiber Communications Conference and Exposition (OFC), San Diego, CA, USA, 11–15 March 2018.
17. Yue, Y.; Wang, Q.; Yao, J.; O'Neil, J.; Pudvay, D.; Anderson, J. 400GbE Technology Demonstration Using CFP8 Pluggable Modules. *Appl. Sci.* **2018**, *8*, 2055. [CrossRef]
18. CFP-MSA. Available online: http://www.cfp-msa.org/ (accessed on 1 March 2019).
19. SFF Committee. Available online: http://www.sffcommittee.com/ie/ (accessed on 1 March 2019).
20. OSFP. Available online: http://osfpmsa.org/index.html (accessed on 1 March 2019).
21. QSFP-DD. Available online: http://www.qsfp-dd.com/ (accessed on 1 March 2019).
22. OIF: Optical Internetworking Forum. Available online: http://www.oiforum.com/ (accessed on 1 March 2019).
23. Cole, C. Future datacenter interfaces based on existing and emerging optics technologies. In Proceedings of the 2013 IEEE Photonics Society Summer Topical Meeting Series, Waikoloa, HI, USA, 8–10 July 2013.
24. Geyer, J.C.; Rasmussen, C.; Shah, B.; Nielsen, T.; Givehchi, M. Power Efficient Coherent Transceivers. In Proceedings of the 2016 European Conference on Optical Communication (ECOC), Dusseldorf, Germany, 18–22 September 2016.
25. Frey, F.; Elschner, R.; Fischer, J.K. Estimation of Trends for Coherent DSP ASIC Power Dissipation for Different Bitrates and Transmission Reaches. In Proceedings of the ITG-Symposium Photonic Networks, Leipzig, Germany, 11–12 May 2017.
26. Zhang, H.; Zhu, B.; Park, S.; Doerr, C.; Aydinlik, M.; Geyer, J.; Pfau, T.; Pendock, G.; Aroca, R.; Liu, F.; et al. Real-time transmission of 16 Tb/s over 1020km using 200Gb/s CFP2-DCO. *Opt. Express* **2018**, *26*, 6943–6948. [CrossRef] [PubMed]
27. Wang, Q.; Yue, Y.; Anderson, J. Compensation of Limited Bandwidth and Nonlinearity for Coherent Transponder. *Appl. Sci.* **2019**, *9*, 1758. [CrossRef]
28. Yue, Y.; Wang, Q.; Anderson, J. Transmitter skew tolerance and spectral efficiency tradeoff in high baud-rate QAM optical communication systems. *Opt. Express* **2018**, *26*, 15045–15058. [CrossRef] [PubMed]
29. Faruk, M.; Savory, S. Digital Signal Processing for Coherent Transceivers Employing Multilevel Formats. *J. Lightwave Technol.* **2017**, *35*, 1125–1141. [CrossRef]

Article

Compensation of Limited Bandwidth and Nonlinearity for Coherent Transponder

Qiang Wang, Yang Yue * and Jon Anderson

Juniper Networks, 1133 Innovation Way, Sunnyvale, CA 94089, USA; qiwang.thresh@gmail.com (Q.W.); jonanderson@juniper.net (J.A.)
* Correspondence: yyue@juniper.net; Tel.: +1-408-745-2000

Received: 27 March 2019; Accepted: 23 April 2019; Published: 28 April 2019

Abstract: Coherent optical transponders are widely deployed in today's long haul and metro optical networks using dense wavelength division multiplexing. To increase the data carrying capacity, the coherent transponder utilizes the high order modulation format and operates at a high baud rate. The limited bandwidth and the nonlinearity are two critical impairments for the coherent in-phase quadrature transmitter. These impairments can be mitigated by digital filters. However, to accurately determine the coefficients of these filters is difficult because the impairment from the limited bandwidth and the impairment from nonlinearity are coupled together. In this paper, we present a novel method to solve this problem. During the initial power-up, we apply a sinusoidal stimulus to the coherent IQ transmitter. We then scan the frequency and amplitude of the stimulus and monitor the output power. By curve-fitting with an accurate mathematical model, we determine the limited bandwidth, the nonlinearity, the power imbalance, and the bias point of the transponder simultaneously. Optimized coefficients of the digital filters are determined accordingly. Furthermore, we utilize a coherent IQ transponder and demonstrate that the limited bandwidth is improved by the finite impulse response filter, while nonlinearity is mitigated by the memoryless Volterra filter.

Keywords: coherent communication; optical communication; pluggable module

---

1. Introduction

Today's telecommunications infrastructure relies on optical fiber communications systems where the coherent in-phase quadrature (IQ) optical transceiver is an essential component. To deliver a large amount of information over a long distance, the high-order quadrature amplitude modulation (QAM) running at a high baud-rate has been adopted in the long-haul transmission system [1,2]. The information is carried on two orthogonal domains. One domain is the in-phase (I) and quadrature (Q), and the other domain is the two orthogonal polarizations, which are X and Y. Thus, four tributary channels are formed, namely XI, XQ, YI, and YQ. Within each tributary, there are impairments due to the limited bandwidth and nonlinearity. Among these tributaries, there are degrading effects due to the time skew and power imbalance. In this work, we mainly focus on the impairments within each tributary.

The limited bandwidth ultimately determines the highest achievable baud rate of the coherent IQ transponder. It is critical to compensate for the limited bandwidth, particularly for the high baud rate coherent IQ transponder. For example, the state-of-art system runs at the all-electronically multiplexed symbol rate of 180 GBd (giga baud) [3]. The nonlinearity causes uneven distribution of the constellation points in the complex plane and eventually limits the signal-to-noise ratio (SNR). Thus, it is equally important to mitigate the nonlinearity for the coherent IQ transponder using the high order QAM. As an example, the next generation coherent DWDM system uses the probability shaping constellation (PSC) QAM [4], and the state-of-the-art demonstration utilizes 4096-PSC-QAM [5].

The limited bandwidth is usually mitigated by the finite impulse response (FIR) filter through the pre-emphasis process in the digital domain. For example, a 61 GBd coherent system can still be demonstrated when the overall electrical bandwidth of the coherent IQ transceiver is less than 15 GHz [6]. To set the FIR filter, one needs to measure the system's bandwidth accurately. A traditional method is to apply a stimulus, for example a sinusoidal signal, to the Mach–Zehnder modulator (MZM). By scanning the frequency of the stimulus and measuring the optical power from the output of MZM, one can determine its bandwidth. To limit the influence of the nonlinearity, the sinusoidal signal has a small amplitude. However, the actual analog signal generated by real data is a large signal applied to the MZM to utilize the dynamic range of the coherent IQ transmitter fully. On the other hand, the bandwidth measured with a large signal is influenced by the nonlinearity, making the measurement result inaccurate.

Multiple methods have been demonstrated to mitigate the nonlinearity. In [7], a lightwave component analyzer is used to determine the limited bandwidth by performing a small-signal measurement. The limited bandwidth is compensated by the first-order kernel of the Volterra filter. Next, the constellation diagram is recovered by the coherent receiver through the digital signal processing (DSP). The second-order kernel and the third-order kernel of the Volterra filters are adaptively updated through the indirect learning algorithm. In such a routine, the nonlinearity is mitigated by minimizing the error function. In [8], the indirect learning algorithm is further expanded so that the limited bandwidth, the time skew between the in-phase tributary and the quadrature tributary, and the nonlinearity are simultaneously compensated by the Volterra filter. In [9], the limited bandwidth, the time skew, and the nonlinearity are simultaneously detected by a high-speed photodiode and a sampling oscilloscope. The high-order Volterra filter is used to compensate those impairments. In [10], a look-up table (LUT) mitigates the nonlinearity and pattern-dependent distortion. The 7-symbol LUT is trained by determining the difference between the training symbol and the actual sample value for the signal. The known symbol sequence is identified by a sliding window, and the address of the LUT is formed accordingly.

Although those developed methods can mitigate the nonlinearity, they rely on the high-speed photodiode to perform optical-to-electrical conversion, and the high-speed digital-to-analog converter (DAC) to perform the measurement. The setup is complicated and cannot be easily integrated within the coherent transponder.

In this article, we demonstrate a novel method to compensate for the limited bandwidth and mitigate the nonlinearity. The most significant advantage of the proposed technique is a simple setup using the low-speed photodiode integrated into the coherent transponder. First, we establish an accurate mathematical model describing the output from the MZM, taking into consideration the limited bandwidth and the nonlinearity. Next, during the initial phase of the coherent transponder, we demonstrate the simultaneous measurement of the limited bandwidth, the nonlinearity, the power imbalance, and the bias point. From the measurement result, we determine the coefficients of the FIR filter and the memoryless Volterra filter. Finally, we compensate for the limited bandwidth and mitigate the nonlinearity of the coherent transmitter.

The article is organized as follows: In Section 2, we establish the mathematic model; in Section 3, we show the measurement result and demonstrate the compensation of the limited bandwidth and the nonlinearity; in Section 4, we discuss multiple aspects of the proposed technique; in Section 5, we draw the conclusions.

The presence of bandwidth limitation and nonlinearity is not unique to the coherent optical transponder. In other types of application, such as radio over fiber [11,12] and millimeter wave band communication [13,14], the transmitter also suffers from the penalty due to bandwidth limitation and nonlinearity. Thus, the novel method demonstrated in this paper can be adopted, modified, and applied to different types of communication systems.

2. Principle

Figure 1 shows a typical coherent IQ transmitter which consists of the digital signal processing (DSP) application specific integrated circuit (ASIC) and the analog coherent optics (ACO). The layer of forward error correction (FEC) adds the overhead error correction. Next, an FIR filter in the tap-and-delay structure compensates the limited bandwidth. The FIR filter is $T_s/2$ spaced, where T_s is the symbol period. A high-speed DAC converts the output of the FIR filter from the digital domain to the analog domain. It is also possible to implement a nonlinear equalizer like the Volterra filter to mitigate the nonlinearity. The analog electrical signal goes through the traces on the radio-frequency (RF) print circuit board (PCB), the pluggable interface (if applicable), the linear RF amplifiers and then are finally applied to the MZM.

The DAC, the RF trace on the PCB, the pluggable connector, the RF electrical amplifier, and the MZM contribute to the limited bandwidth. The intrinsic transfer function of MZM in the form of sinusoidal function and the nonlinear amplitude response within the data path contribute to the nonlinearity.

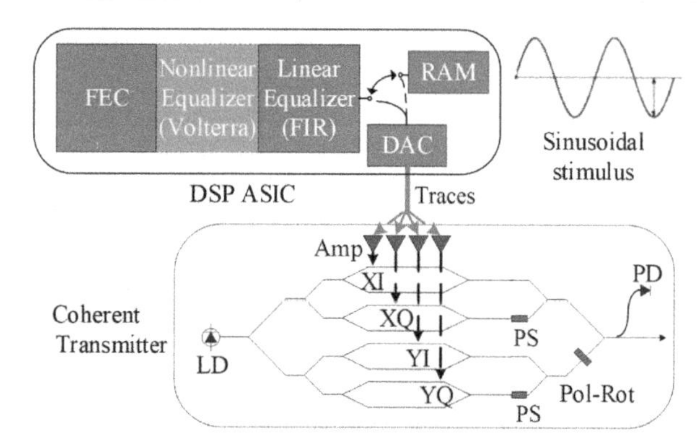

Figure 1. The block diagram of the coherent IQ transmitter and the DSP ASIC. PS: phase shifter, Pol-Rot: polarization rotator. DAC input can be switched between RAM and regular path.

Also, an onboard random-access memory (RAM) can be implemented in the DSP ASIC for testing and diagnosis. During the initial calibration, one can load RAM with the desired data pattern, and set the output of the DAC according to the content in the RAM. After the initial calibration, one can switch back to the regular data path. We utilize this feature to demonstrate our technique. Initially, during the power-up, we load the RAM with a sinusoidal stimulus to one tributary at a time and write 0 to the other tributaries. Thus, the output is from the tributary with the stimulus. Then, we can apply the following equation

$$P_{out} = P_{stdy} \cos^2\left(0.5\pi\left(V_{swing}/V_\pi + V_{bias}/V_{null}\right)\right), \tag{1}$$

here P_{out} is the output from the IQ transmitter, monitored by a low-speed photodiode (PD). P_{stdy} is the steady-state power of the tributary under stimulus, V_{swing} is the voltage applied to the MZM, V_π is the voltage achieving π phase shift, V_{bias} is the bias voltage applied to MZM, V_{null} is the bias voltage required for the null point (corresponding to $\pi/2$ phase shift), $cos^2()$ is the intrinsic transfer function of the MZM.

V_{DAC} is the maximum output voltage of the DAC, IL_{RF} is the insertion loss of the RF traces, $Gain_{AMP}$ is the gain of the RF amplifier, BW_{MZM} is the bandwidth of the MZM. The parameters above are frequency-dependent and thus contribute to the limited bandwidth. N_{sig} and ω are the amplitude and frequency of the sinusoidal stimulus. Bit_{DAC} is the number of bits of the high-speed DAC, where the first bit is a sign bit. Consequently, V_{swing} can be expressed as the following:

$$V_{swing} = N_{sig}\sin(\omega t) * V_{DAC}(\omega) * IL_{RF}(\omega) \\ *Gain_{AMP}(\omega) * BW_{MZM}(\omega)/2^{(Bit_{DAC}-1)} \quad , \tag{2}$$

Furthermore, we define the normalized signal amplitude x, the bandwidth factor α, the bias factor β, and the MZM phase shift φ_{MZM}. The output from the MZM is expressed as

$$\begin{aligned} & x = N_{sig}/2^{(Bit_{DAC}-1)}, x \in [0,1), \\ & \alpha_\omega = V_{DAC} * IL_{RF} * Gain_{AMP} * BW_{MZM}/V_\pi, \\ & \beta = V_{bias}/V_{null}, \\ & \phi_{mzm} = 0.5\pi\alpha_\omega x \sin(\omega t), \\ & P_{out}(\omega, x) = P_{stdy}\cos^2(\phi_{mzm} + 0.5\pi\beta) \end{aligned} \quad , \tag{3}$$

The nonlinear amplitude response is not included in Equation (3). In [15,16], the nonlinear response is treated as a quadrature term in x. Adding nonlinear response, φ_{mzm} is expressed as

$$\phi_{mzm} = 0.5\pi\alpha_\omega\left(\gamma_\omega x^2 + x\right)\sin(\omega t), \tag{4}$$

here γ is the coefficient for a nonlinear response. The subscript of α and γ indicates that they are dependent on the frequency.

Using the Jacobi–Anger expansion [17], one can show that Equation (4) can be written as the following

$$\begin{aligned} & P_{out}(\omega, x) = 0.5P_{stdy} + 0.5P_{stdy}\cos(\pi\beta)J_0\left(\pi\alpha_\omega\left(\gamma_\omega x^2 + x\right)\right) \\ & + P_{stdy}\cos(\pi\beta)\sum_{m=1}^{\infty} J_{2m}\left(\pi\alpha_\omega\left(\gamma_\omega x^2 + x\right)\right)\cos(2m\omega t) \\ & - P_{stdy}\sin(\pi\beta)\sum_{m=1}^{\infty} J_{2m-1}\left(\pi\alpha_\omega\left(\gamma_\omega x^2 + x\right)\right)\sin((2m-1)\omega t) \end{aligned} \quad , \tag{5}$$

here, $J_m(\,)$ is the first-kind m-th Bessel function. The average output power over the time $P_{avg}(\omega,x)$, detected by a low-speed PD, can be expressed as

$$P_{avg}(\omega, x) = 0.5P_{stdy}\left[1 + \cos(\pi\beta)J_0\left(\pi\alpha_\omega\left(\gamma_\omega x^2 + x\right)\right)\right], \tag{6}$$

We fix ω and scan the amplitude of the sinusoidal stimulus between 0 and 2^(Bit_{ADC}-1). We get a curve of $P_{avg}(\omega,x)$ versus x. The underlying fitting parameter [P_{stdy}, α_ω, β, γ_ω] can be extracted using a curve fitting method like sequential quadratic programming (SQP). The SQP minimizes the relative root-mean-square (RMS) error

$$Err_{rms} = \frac{\sum_{k=1}^{K}\left[P_{avg}^{Meas}(x_k) - P_{avg}^{Fit}(x_k)\right]^2}{K\left[P_{avg}^{Meas}(x_k)\right]^2}, \tag{7}$$

where superscript "*Meas*" and "*Fit*" indicate the measurement results and the fitting results. Here K and k are the total measurements and the index of measurement.

Next, we scan the frequency from a value close to DC (for example, a few GHz) to the value of the baud rate. We record the optical power from each tributary when the amplitude of the sinusoidal stimulus is varied from zero to full swing. We perform the curve fitting according to Equations (6) and (7). Then, we determine the α_ω and γ_ω over the different frequencies. Accordingly, we find the coefficients of the FIR filter and Volterra filter to mitigate the limited bandwidth and the nonlinearity. Furthermore, the power imbalance between tributaries and the bias point of the MZM can be decided by P_{stdy} and β. The DAC's input returns to the regular data path for normal operation after the impairments are calibrated.

3. Experimental Results

3.1. Measured Bandwidth and Nonlinearity

The proposed technique is validated on a pluggable CFP2-ACO module (CFP2 form factor analog coherent optics) and a DSP ASIC [18]. The pluggable module is a class-2 CFP2-ACO with a linear RF amplifier, which can support 200 Gb/s (gigabit per second) traffic using 16-QAM or 300 Gb/s traffic using 64-QAM, at the baud rate of ~30 GBd. The DSP ASIC has the built-in RAM, allowing the scanning of sinusoidal amplitude and frequency as discussed above. To minimize any interruption during the initial calibration, the gain of the RF amplifier and the bias point of MZM remain constant.

Figure 2 shows the results over four tributaries with ω = 3.77 GHz. Table 1 shows the fitting parameters and relative RMS error. The fitting curves and the measured curves almost overlap, which indicates that the proposed model in Equation (6) is very accurate. In addition, the power imbalance between tributaries and the bias point are determined accordingly.

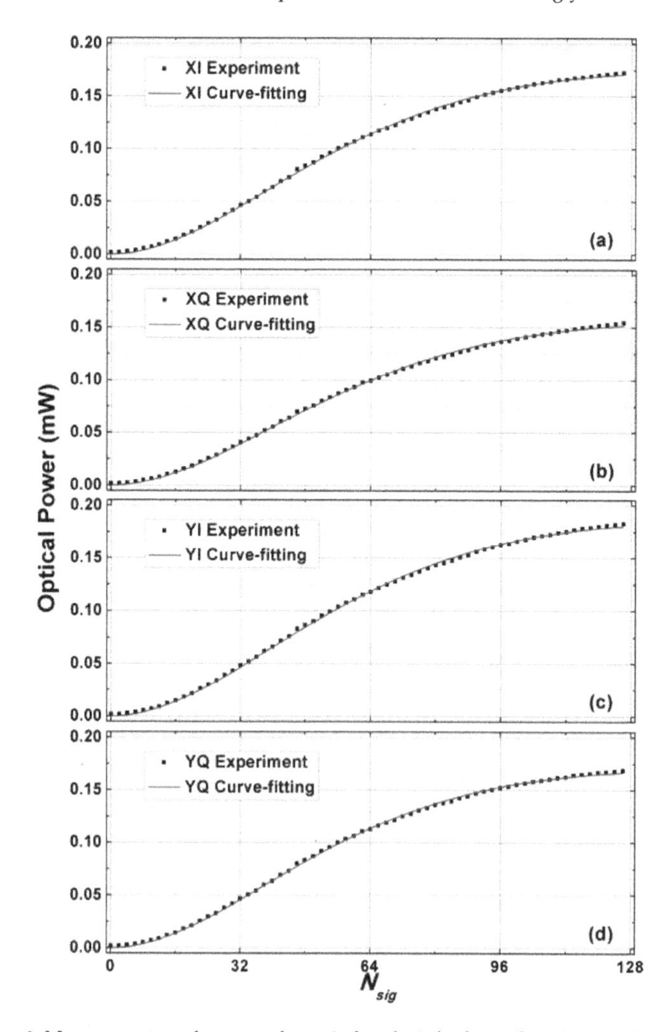

Figure 2. Measurement results versus theoretical analysis for four tributaries. ω = 3.77 GHz.

Table 1. Fittings parameters for four tributaries.

Parameters [1]	XI Trib	XQ Trib	YI Trib	YQ Trib
$P_{stdy}(mW)$	0.2712	0.2423	0.2891	0.2613
α	1.7396	1.7042	1.6991	1.7803
β	1.0000	1.0000	1.0000	1.0000
γ	−0.4503	−0.4421	−0.4429	−0.4542
Err_{rms}	4.02×10^{-2}	4.18×10^{-2}	3.95×10^{-2}	3.98×10^{-2}

[1] ω = 3.77 GHz.

Next, we scan the frequency of the sinusoidal stimulus, repeat the measurement and extract the fitting parameters. The purpose is to determine those parameters' dependency on the frequency component. Ideally, we can also extract four fitting parameters from the measurement curve. However, the output power from MZM is quite low at the high frequency close to the baud rate due to the limited bandwidth. We experience a large fitting error if we use four fitting parameters. As the steady-state power remains relatively unchanged, and the bias point has a negligible drift over time, we can assume that P_{stdy} and β remain unchanged over the different frequencies. Thus, we only extract four fitting parameters from the measurement curve using a sinusoidal stimulus running at a low-frequency ω, for example at 3.77 GHz as shown in Figure 2. We then apply the same P_{stdy} and β values to all other measurement curves at different frequencies. We extracted the fitting parameters α and γ for various frequencies. This improves the fitting error, particularly when ω is close to the baud rate. Over the different frequency, the model is accurate as shown in Figure 3.

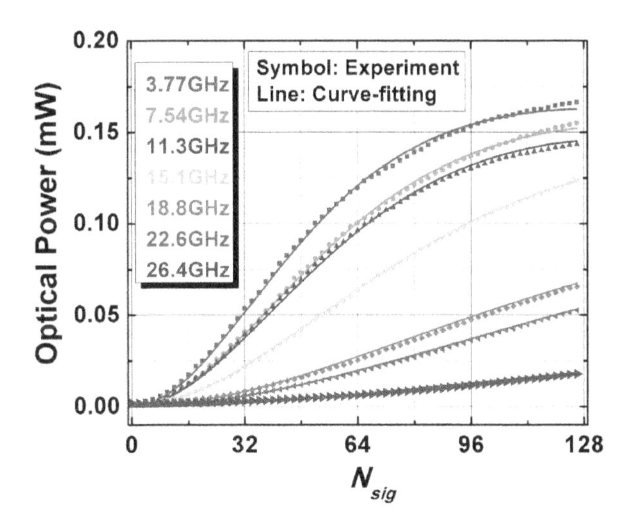

Figure 3. Results over different frequencies. Measurement results agree well with the theoretical analysis.

Then, we determine the limited bandwidth by dividing the α value at a higher frequency with the α value at 3.77 GHz, which is low enough to be used as the reference point. The limited bandwidth α is plotted in Figure 4 on the left axis. We also calculate the limited bandwidth determined through traditional power measurement. It is done through the following steps: one fixes N_{sig}, then scans ω, and measures the average output power P_{avg}. Then, the P_{avg} at the higher frequency is divided by the P_{avg} close to the DC frequency to determine the limited bandwidth. It is noticeable that both the small-signal result (for example, N_{sig} = 32) and the large-signal result (for example, N_{sig} = 127) show the deviation against the actual bandwidth (α curve). The reason is that the nonlinear response causes the actual phase shift applied to MZM to be smaller than the intended value.

The nonlinear response γ is plotted in Figure 4 towards the right axis. One can notice that within the Nyquist frequency (Ω_{NY}), the dependency of γ on ω is small. In addition, all frequency components within Ω_{NY} will have roughly the same amplitude when the limited bandwidth is compensated. Thus, we can define the average nonlinear response γ_{avg} within the Nyquist frequency. Since Nyquist pulse shaping is widely used to improve the spectral efficiency, we ignore the frequency components outside Ω_{NY} which are eliminated by the sharp roll-off of the Nyquist filter.

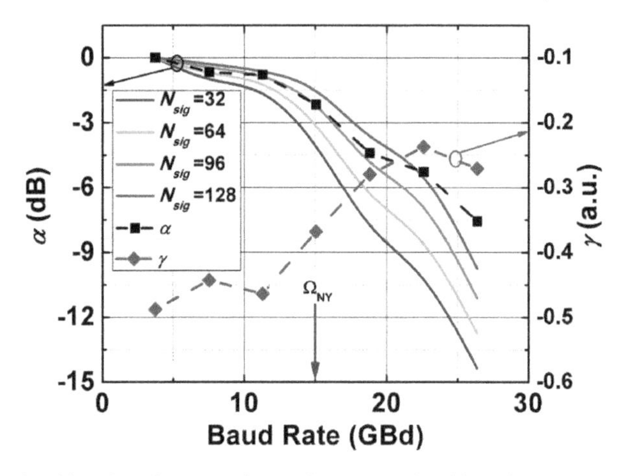

Figure 4. Bandwidth and nonlinearity of one tributary. Bandwidth is plotted on the left axis and nonlinearity is plotted on the right axis.

To mitigate the nonlinear response, a quadratic equation is first solved. There are two roots for the quadratic equation. We can choose the correct root because the normalized amplitude x is between 0 and 1. Furthermore, to obtain the coefficients of the Volterra filter for compensation of nonlinearity, a Taylor expansion is performed around the zero point, as shown below:

$$\begin{aligned} y &= \gamma_{avg}x^2 + x \\ x &= \left(\sqrt{1+4\gamma_{avg}y} - 1\right)/2\gamma_{avg} \approx y - \gamma_{avg}y^2 + 2\gamma_{avg}^2 y^3 \end{aligned}, \quad (8)$$

As seen, the coefficients of a memoryless Volterra filter are exactly the coefficients of a Taylor expansion. Thus, a third-order memoryless Volterra filter can be implemented in DSP ASIC to reverse the nonlinear response. The coefficients of the Volterra filter are simple and directly correlated to nonlinear response γ_{avg}. Furthermore, MZM's transfer function is a sinusoidal function. Its compensation function is *"arcsin"* function which can be realized through a LUT in the DSP ASIC.

As discussed in [7], the coherent transmitter can be modeled as a Wiener system, where the static nonlinear response follows the dynamic linear response. To compensate for the Wiener system, a Hammerstein system is used, where the static nonlinear equalization is placed in front of the dynamic linear equalization [19].

Figure 5 shows the block diagram of the proposed DSP architecture. The digital signal is first pre-distorted by the third-order memoryless Volterra filter to mitigate the nonlinearity. Then, the sinusoidal transfer function of the MZM is compensated by the *"arcsin"* LUT. At last, the limited bandwidth is compensated through the pre-emphasis implemented by the FIR filter. The tap coefficients of the memoryless Volterra filter is decided by γ_{avg} and the tap coefficients of the FIR filter is set such that the frequency response of the FIR filter is inversely proportional to the limited bandwidth (α).

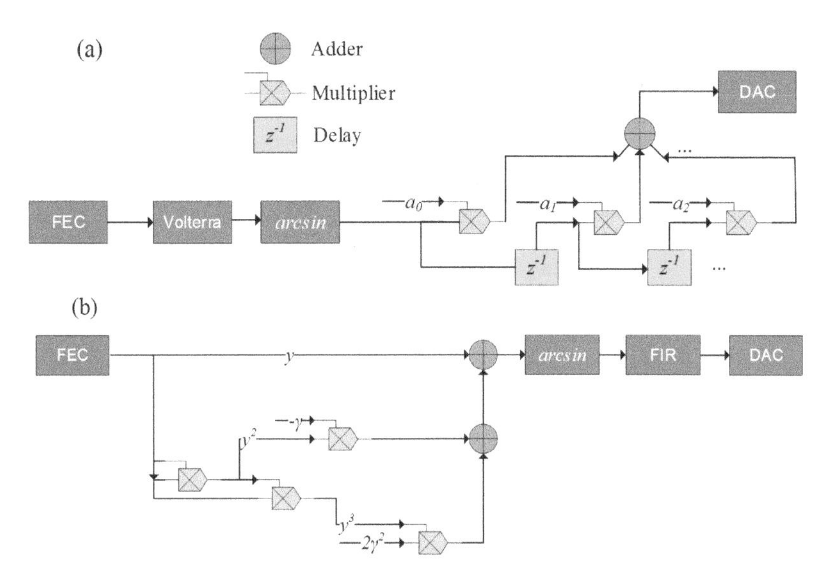

Figure 5. Proposed DSP block diagram. (**a**) The proposed structure of the FIR filter is shown in detail. a_0, a_1, a_2, and the rest are the tap coefficients determined from α. (**b**) The proposed structure of the Volterra filter is shown in detail.

3.2. Compensation for Bandwidth and Nonlinearity

To demonstrate this technique, we generate 300 Gb/s signal using 64-QAM format at approximately 30 GBd. This maximum baud rate is limited by our DSP ASIC. The DSP ASIC currently has no built-in Volterra filter. Thus, our proposed DSP architecture cannot be realized directly through the data path. So, the pseudo-random binary sequence (PRBS) is first generated. Then, the Gray coding is applied to map the QAM signal so that there is only a one-bit difference between the adjacent constellation points. Next, the memoryless Volterra filter and the '*arcsin*' function are applied to the signal. In the end, the signal is convoluted with an FIR filter. There are multiple functionalities of the FIR filter: to overcome the limited bandwidth α coming from the components on the data path; to perform Nyquist pulse shaping using root raised cosine (RRC) filter with the roll-off factor of 0.1 [20]; to compensate the skew among the tributaries [21]. The data stream is loaded into the on-chip RAM and the output of the DAC is set according to the data in the RAM.

The coherent transmitter then converts the electrical signal to an optical signal. The optical output from the coherent transmitter is further captured by a coherent receiver, which consists of 90-degree optical hybrid, balanced photodiodes, and linear trans-impedance amplifiers. An integrated tunable laser assembly (ITLA) serves as the local oscillator (LO). The electrical signal is then converted from the analog domain to the digital domain through a real-time oscilloscope (80 G samples per second, 33 GHz bandwidth).

The offline DSP is used to generate a constellation diagram and count the bit error rate. The DSP processing steps are similar to those described in [22]. First, a static equalization with a matched RRC filter is implemented in the time domain. Second, the chromatic dispersion is compensated in the frequency domain. Next, the polarization de-multiplexing and the compensation of polarization mode dispersion are implemented through an adaptive equalizer using the radius directed equalization (RDE) [23]. Then in the following steps, the timing is recovered using Gardner's method, the frequency offset is estimated, the carrier phase is recovered, and the symbol is determined. The detailed diagram for the experimental setup is shown in Figure 6.

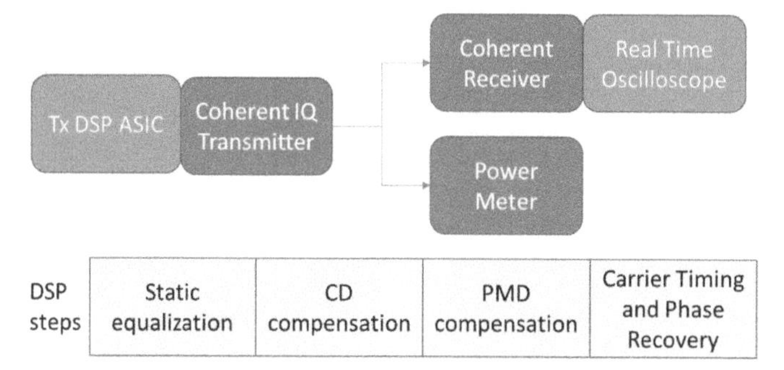

Figure 6. Experimental setup. A power meter is used to perform initial calibration with sinusoidal stimulus. Coherent receiver and real-time oscilloscope are used to recover the constellation diagram. The processing steps of DSP are also shown.

The constellation diagram can only be recovered when the limited bandwidth is compensated by the FIR filter. We compare the constellation diagram before and after the compensation of the nonlinearity, as shown in Figure 7. The gain of RF amplifier is set to a high value so that MZM is driven into a nonlinear region. As seen, without the compensation of nonlinearity, the constellation points are crowded and blurry. After the nonlinearity is mitigated, the constellation points are linearly distributed and separated.

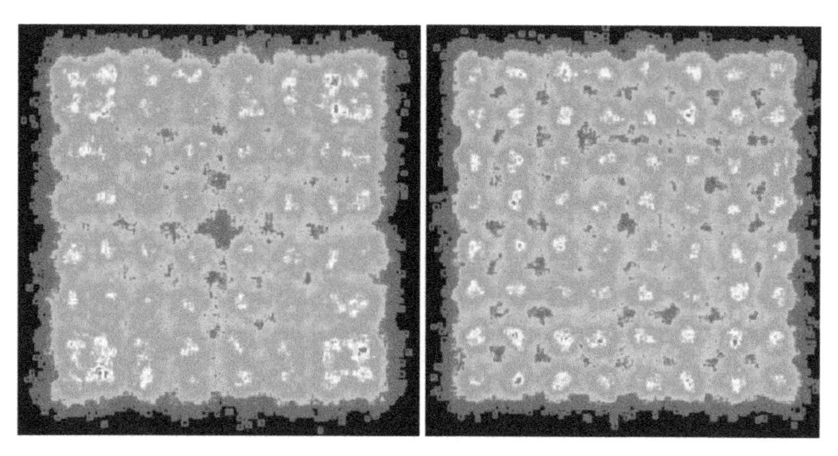

Figure 7. Constellation diagram without Volterra filter (**left**) and that with Volterra filter (**right**). Both results are obtained with bandwidth compensation and arcsin compensation.

The improvement in Q^2 factor is approximately 0.3 dB. The improvement is smaller than expected due to the hardware implementation penalty at 64-QAM. In our setup, the implementation penalty is higher than expected, mainly due to the limit effective number of bit (ENOB) in DAC and ADC. The implementation penalty acts as a noise floor for our system. The contribution from the nonlinearity is small compared with the implementation penalty. After removing the nonlinearity, the performance of the coherent transponder is still dominated by the implementation penalty. Thus, the improvement from the mitigation of the nonlinearity is smaller than expected.

4. Discussion

During the measurement of limited bandwidth and nonlinearity, we adjust the bias voltage so that the MZM is biased at its null point. The optimal bias voltage can be obtained by turning off the modulation signal and scanning the bias voltage. The bias voltage resulting in the minimum output power is the optimal voltage. The coherent IQ transmitter in our experiment is fabricated in Indium Phosphide (InP) material and its bias point is stable during the measurement. As seen from Table 1, the measured bias point factor is close to 1, indicating that the bias point is stably held at the optimal null point. During the measurement, the automatic bias control (ABC) circuit, which is based on the dithering of bias voltage, is turned off to avoid any potential interference. During the normal operation, the ABC circuit is enabled so that any long-term drift can be compensated.

In addition to the limited bandwidth (α) and the nonlinearity (γ), our method can also measure the power imbalance (P_{stdy}) among tributaries and the bias setting point (β). To achieve optimal performance, the power imbalance among the tributaries needs to be minimized. Our method offers a novel way to measure the power imbalance. Once measured, the power imbalance can be compensated by adjusting the variable optical attenuator or the semiconductor optical amplifier which can be integrated into the coherent transmitter [24]. The bias setting can be also optimized by adjusting the bias voltage applied to the MZM.

The frequency and the amplitude of the sinusoidal stimulus are controlled in the digital domain. Thus, the sinusoidal stimulus is generated with high accuracy. N_{sig} and ω can be scanned with a fine step, and the corresponding P_{avg} can be measured. It is well known that by measuring multiple data points and applying the curve fitting, the measurement accuracy can be greatly improved. Assuming K as the total measurement points, the improvement in the accuracy is proportional to the square root of K. In this way, we can significantly improve the measurement accuracy of the limited bandwidth, the nonlinearity, the power imbalance and the bias point among the four tributaries.

A Volterra filter which has multiple taps (memory) in the high-order kernels can mitigate the frequency-dependent nonlinear effect. In [7], the memory depth of the second-order kernel is 7 and the memory depth of the third order kernel is 7. Note that the frequency-dependent nonlinear response is also captured during our measurement. Thus, the coefficients of the Volterra filter with the memory can be determined accordingly. However, the Volterra filter with the memory has higher complexity, larger power consumption, and greater latency. The trade-off between performance and complexity should be carefully considered. In this work, we utilize the memoryless Volterra filter to mitigate the average nonlinearity for its simplicity. Thus, the frequency-dependent component of the nonlinearity is neglected. Still, a noticeable improvement was observed after the nonlinearity compensation was realized, particularly for MZM working in the nonlinear regime.

5. Conclusions

Next-generation coherent transponders operate at high baud rate and utilize advanced QAM modulation. For these transponders, the limited bandwidth and the nonlinearity are two severe impairments which need to be mitigated. The limited bandwidth is mitigated by the FIR filter and the nonlinearity is mitigated by the memoryless Volterra filter for its simplicity.

A novel technique to determine the tap coefficients for the FIR filter and the memoryless Volterra filter is presented. A sinusoidal stimulus is applied during the initial power-up. We then scan the amplitude and frequency of the sinusoidal stimulus and monitor the output from the coherent transmitter. Then, we use the curve fitting method to determine the underlying parameters of the coherent transmitter, such as the limited bandwidth, the nonlinearity, the power imbalance among the tributaries, and the bias point. Accordingly, we determine the tap coefficients of the FIR filter and the tap coefficients of the Volterra filter.

We apply this technique on a DSP ASIC and a coherent CFP2-ACO transponder. We drive the coherent IQ transponder into the highly nonlinear regime. When the nonlinearity is mitigated by the memoryless Volterra filter, we achieve a noticeable improvement in the constellation diagram. After the compensation of the limited bandwidth and the nonlinearity, we demonstrate a coherent IQ transponder with 300 Gb/s data rate using the 64-QAM modulation format and running at the 30 GBd baud rate.

Author Contributions: Conceptualization, methodology, software, validation, formal analysis, investigation, resources, data curation, writing—original draft preparation, writing—review and editing, visualization: Q.W. Conceptualization, methodology, software, validation, formal analysis, investigation, resources, data curation, writing—original draft preparation, writing—review and editing, visualization: Y.Y. Resources, writing—review and editing, supervision: J.A.

Funding: This research received no external funding.

Acknowledgments: The authors also gratefully acknowledge vigorous encouragement and sturdy support on innovation from Domenico Di Mola at Juniper Networks.

Conflicts of Interest: The authors declare no conflicts of interest.

References

1. Roberts, K.; Zhuge, Q.; Monga, I.; Gareau, S.; Laperle, C. Beyond 100 Gb/s: Capacity, flexibility, and network optimization. *J. Opt. Commun. Netw.* **2017**, *9*, c12–c24. [CrossRef]
2. Wang, Q.; Yue, Y.; He, X.; Vovan, A.; Anderson, A. Accurate model to predict performance of coherent optical transponder for high baud rate and advanced modulation format. *Opt. Express* **2018**, *26*, 12970–12984. [CrossRef]
3. Buchali, F.; Steiner, F.; Böcherer, G.; Schmalen, L.; Schulte, P.; Idler, W. Rate Adaptation and Reach Increase by Probabilistically Shaped 64-QAM: An Experimental Demonstration. *J. Lightwave Technol.* **2016**, *34*, 1599–1609. [CrossRef]
4. Raybon, G.; Adamiecki, A.; Cho, J.; Jorge, F.; Konczykowska, A.; Riet, M.; Duval, B.; Dupuy, J.Y.; Fontaine, N.; Winzer, P.J. 180-GBaud All-ETDM Single-Carrier Polarization Multiplexed QPSK Transmission over 4480 km. In Proceedings of the Optical Fiber Communication Conference, San Diego, CA, USA, 11–15 March 2018. paper Th4C.3.
5. Olsson, S.L.I.; Cho, J.; Chandrasekhar, S.; Chen, X.; Winzer, P.J.; Makovejs, S. Probabilistically shaped PDM 4096-QAM transmission over up to 200 km of fiber using standard intradyne detection. *Opt. Express* **2018**, *26*, 4522–4530. [CrossRef]
6. Zhang, Z.; Li, C.; Chen, J.; Ding, T.; Wang, Y.; Xiang, H.; Xiao, Z.; Li, L.; Si, M.; Cui, X. Coherent transceiver operating at 61-Gbaud/s. *Opt. Express* **2014**, *23*, 18988–18995. [CrossRef]
7. Berenguer, P.W.; Nölle, M.; Molle, L.; Raman, T.; Napoli, A.; Schubert, C.; Fischer, J.K. Nonlinear Digital Pre-distortion of Transmitter Components. *J. Lightwave Technol.* **2016**, *34*, 1739–1745. [CrossRef]
8. Khanna, G.; Spinnler, B.; Calabrò, S.; Man, E.D.; Hanik, A. A Robust Adaptive Pre-Distortion Method for Optical Communication Transmitters. *IEEE Photonics Technol. Lett.* **2016**, *28*, 752–755. [CrossRef]
9. Duthel, T.; Hermann, P.; Schiessl, J.; Fludger, C.R.S.; Bisplinghoff, C.; Kupfer, T. Characterization and Pre-Distortion of Linear and Non-Linear Transmitter Impairments for PM-64QAM Applications. In Proceedings of the European Conference and Exhibition on Optical Communications, Düsseldorf, Germany, 18–22 September 2016. paper W.4.P1.SC3.6.
10. Ke, J.; Gao, Y.; Cartledge, G.C. 400 Gbit/s single-carrier and 1 Tbit/s three-carrier super channel signals using dual polarization 16-QAM with look-up table correction and optical pulse shaping. *Opt. Express* **2014**, *22*, 71–83. [CrossRef] [PubMed]
11. Amir, I.; Bunruangses, M.; Chaiwong, K.; Udaiyakumar, R.; Maheswar, R.; Hindia, M.N.; Dimyati, K.B.; Yupapin, P. Dual-wavelength transmission system using double micro-resonator system for EMI healthcare applications. *Microsyst. Technol.* **2018**, *25*, 1185–1193. [CrossRef]
12. Amir, I.; Hindia, M.N.; Reza, A.W.; Ahmad, H. LTE smart grid performance gains with additional remote antenna units via radio over fiber using a microring resonator system. *Opt. Switch. Netw.* **2017**, *25*, 13–23. [CrossRef]
13. Hindia, M.; Qamar, F.; Rahman, T.A.; Amiri, I.S. A stochastic geometrical approach for full-duplex MIMO relaying model of high density network. *Ad Hoc Netw.* **2018**, *74*, 34–36. [CrossRef]
14. Qamar, F.; Hindia, M.N.; Abbas, T.; Dimyati, K.B.; Amiri, I.S. Investigation of QoS Performance Evaluation over 5G Network for Indoor Environment at millimeter wave Bands. *Int. J. Electron. Telecommun.* **2019**, *65*, 95–101.
15. Li, Y.; Wang, R.; Bhardwaj, A.; Ristic, S.; Bowers, J. High Linearity InP-Based Phase Modulators Using a Shallow Quantum-Well Design. *IEEE Photonics Technol. Lett.* **2010**, *22*, 1340–1342. [CrossRef]

16. Zhou, Y.; Zhou, L.; Su, F.; Li, X.; Chen, J. Linearity Measurement and Pulse Amplitude Modulation in a Silicon Single-Drive Push–Pull Mach–Zehnder Modulator. *J. Lightwave Technol.* **2016**, *34*, 3323–3329. [CrossRef]
17. Abramowitz, M.; Stegun, I. *Handbook of Mathematical Functions with Formulas, Graphs, and Mathematical Tables*; Applied Mathematics Series; Department of Commerce, National Institute of Standards and Technology: Gaithersburg, ML, USA, 1972; p. 55.
18. Lu, F.; Zhang, B.; Yue, Y.; Anderson, J.; Chang, G.-K. Investigation of Pre-Equalization Technique for Pluggable CFP2-ACO Transceivers in Beyond 100 Gb/s Transmissions. *J. Lightwave Technol.* **2017**, *35*, 230–237. [CrossRef]
19. Pan, J.; Cheng, C.-H. Wiener–Hammerstein Model Based Electrical Equalizer for Optical Communication Systems. *J. Lightwave Technol.* **2011**, *29*, 2454–2459. [CrossRef]
20. Yue, Y.; Wang, Q.; Anderson, J. Transmitter skew tolerance and spectral efficiency tradeoff in high baud-rate QAM optical communication systems. *Opt. Express* **2018**, *26*, 15045–15058. [CrossRef]
21. Yue, Y.; Zhang, B.; Wang, Q.; Lofland, R.; O'Neil, J.; Anderson, J. Detection and alignment of dual-polarization optical quadrature amplitude transmitter IQ and XY skews using reconfigurable interference. *Opt. Express* **2016**, *24*, 6719–6734. [CrossRef]
22. Faruk, M.S.; Savory, S.J. Digital Signal Processing for Coherent Transceivers Employing Multilevel Formats. *J. Lightwave Technol.* **2017**, *35*, 1125–1141. [CrossRef]
23. Fatadin, I.; Ives, D.; Savory, S.J. Blind Equalization and Carrier Phase Recovery in a 16-QAM Optical Coherent System. *J. Lightwave Technol.* **2009**, *27*, 3042–3049. [CrossRef]
24. Wang, Q.; Yue, Y.; Anderson, J. Detection and compensation of power imbalance, modulation strength, and bias drift in coherent IQ transmitter through digital filter. *Opt. Express* **2018**, *26*, 23069–23083. [CrossRef]

Article

Optical Transmitters without Driver Amplifiers—Optimal Operation Conditions

Arne Josten *, Benedikt Baeuerle, Romain Bonjour, Wolfgang Heni and Juerg Leuthold

Institute of Electromagnetic Fields (IEF), ETH Zurich, 8092 Zurich, Switzerland; bbaeuerle@ethz.ch (B.B.); rbonjour@ethz.ch (R.B.); wheni@ethz.ch (W.H.); leuthold@ethz.ch (J.L.)

* Correspondence: ajosten@ethz.ch; Tel.: +41-446-325-224

Received: 22 August 2018; Accepted: 10 September 2018; Published: 14 September 2018

Abstract: An important challenge in optical communications is the generation of highest-quality waveforms with a Mach–Zehnder modulator with a limited electrical swing (V_{pp}). For this, we discuss, under limited V_{pp}, the influence of the waveform design on the root-mean-square amplitude, and thus, the optical signal quality. We discuss the influence of the pulse shape, clipping, and digital pre-distortion on the signal quality after the electrical-to-optical conversion. Our simulations and experiments, e.g., suggest that pre-distortion comes at the expense of electrical swing of the eye-opening and results in a lower optical signal-to-noise ratio (OSNR). Conversely, digital post-distortion provides operation with larger eye-openings, and therefore, provides an SNR increase of at least 0.5 dB. Furthermore, we find that increasing the roll-off factor increases the electrical swing of the eye-opening. However, there is negligible benefit of increasing the roll-off factor of square-root-raised-cosine pulse shaped signals beyond 0.4. The findings are of interest for single-channel intensity modulation and direct detection (IM/DD) links, as well as optical coherent communication links.

Keywords: optical communications; fiber optics communications; modulators; mitigation of optical transceiver impairments

1. Introduction

Low power consumption is one of the major design goals for the next-generation highly integrated datacom systems [1]. The power consumption of an optical transmitter can be significantly reduced by driver-amplifier-less operation [2]. In transmitters without a driver amplifier, the electrical source is directly connected to the electro-optical modulator [3–5]. This reduces the transmitter power consumption [2], improves the link-noise figure [6], and avoids additional noise and non-linear distortions [7,8]. Driver-amplifier-less optical transmitters generating multilevel signals were even shown without digital-to-analog converters (DACs) [9,10]. A key enabler for the driver-amplifier-less optical transmitter technology is an optical modulator with a small V_π (voltage to switch the modulator from on to off). In fact, numerous optical modulators that can be operated with driving voltages below 1 V_{pp} (electrical swing) were demonstrated [3,11–15] with speeds up to 100 GBd [16]. For the best performance, the frequency response of optical transmitters must be corrected to mitigate distortions and the resulting inter-symbol interference (ISI). However, the compensation of the transmitter frequency response is known to minimize the electrical output signal power of the DAC [17]; this is due to the increase in the peak-to-average power ratio (PAPR) in pre-distorted signals [18–20]. A loss in the electrical signal power is a particular challenge for driver-amplifier-less optical transmitters, because they already have a reduced margin concerning electrical signal power. Therefore, the correct design of the waveform, and eventually, a wisely chosen clipping [21–27] are required.

State-of-the-art optical transmitters circumvent the penalty of loss in the electrical signal due to predistortion by using electrical driver amplifiers. These amplifiers boost the electrical signal to allow a driving of the modulators in the range of V_π. Preamplifiers give sufficient margin to apply a digital pre-distortion (DPreD). In fact, digital pre-distortion techniques are more and more deployed in commercial systems. Thus, for instance, a low-cost self-calibration routine to determine the frequency response including the in-phase and quadrature (IQ) skew [28] was developed, and real-time transmission of 1 Tb/s with a 28-nm CMOS ASIC digitally compensating for up to 25 dB of loss in electrical-to-optical (EO) and optical-to-electrical (OE) conversion [29] was shown. DPreD can also cancel, in part, the linear and non-linear distortions. The linear pre-distortion methods compensate for the low-pass characteristic of the transmitter, and therefore, eliminate the inter-symbol interference (ISI) of the signal [30–37]. Non-linear compensation methods can cancel the non-linear transfer function of driver amplifiers, digital-to-analog converters, and Mach–Zehnder modulators. In the case of a dominant non-linear influence on the signal, either Volterra series equalization [38–40] or pattern-dependent look-up table-based corrections [41,42] were demonstrated to increase the system performance. Furthermore, the nonlinear Mach–Zehnder modulator transfer function can, for instance, also be linearized by inversion of the latter [40,43] or by the design of a special modulator [44].

However, because of an increasing demand for transmitters with a reduced power consumption and the emergence of high-speed optical modulators with a small V_π, it is of interest how to design a waveform that optimizes the optical signal-to-noise ratio (OSNR) generated by a driver-amplifier-less optical transmitter.

In this paper, we extend our previous work [45] and show that the correct choice of roll-off factor (ROF) for the signal-pulse shape in combination with clipping and that the digital post-distortion (DPostD), rather than DPreD, enhance the optical signal quality of the driver-amplifier-less transmitters. More precisely, we show that increasing the ROF of a square-root-raised-cosine (SRRC) signal results in a larger electrical eye-opening; however, increasing the ROF beyond 0.4 does not bring benefit with respect to electrical eye-opening. Furthermore, we investigate the impact of clipping to SRRC signals. We also report that DPreD reduces the root-mean square (RMS) amplitude. Instead, DPostD, rather than DPreD, can be used to correct for the transmitter frequency response without influencing the transmitter electrical signal power, and subsequently, the OSNR after the optical modulator. Our simulative analyses were obtained with MATLAB and Mathematica. We substantiated our findings with measurement results showing the gain by DPostD in dependence of DAC output swing, roll-off factor, and symbol rate.

2. Influence of Pulse Shape on the Electrical Eye-Opening

The swing of the electrical eye-opening depends on the maximum available DAC output voltage and the chosen pulse shape. Generally, the maximum of the signal waveform is normalized to the maximum of the DAC output swing d_{DAC}. The eye-opening d_{Eye} is then a fraction thereof, as shown Figure 1. This fraction depends on the pulse shape.

In the past, pulse shaping was mostly done by pulse carving. Here, a second optical modulator was used to shape the form of the optical pulses in a desired manner [46–49]. Pulse carving with different duty cycles showed advantages in transmission distance [50] and also enabled high-capacity links [51]. Carving of sinc-shaped pulses with very high quality was also shown [52]. However, due to the simplified set-up and versatility, we restricted the analysis of this article to digital pulse shaping and the use of only one optical modulator. Digital pulse shaping is extremely versatile and allows for arbitrary pulse shapes. Due to the advantageous characteristics in transmission, raised-cosine pulse shapes emerged. They are an efficient trade-off between the rectangular and the sinc-like pulse. In transceivers, the raised-cosine (RC) frequency response can be split into a square-root-raised-cosine (SRRC) filter at the transmitter and an SRRC filter at the receiver. This allows matched filtering and leads to inter-symbol interference (ISI)-free waveforms [53]. The excess bandwidth B of these filters depends on the roll-off factor (ROF) and the symbol rate R, and is calculated by $B = 0.5(1 + ROF)R$.

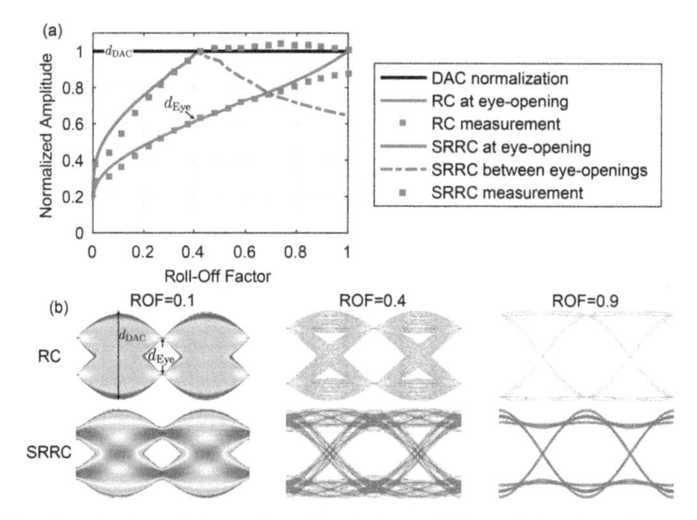

Figure 1. Investigation of transmitter electrical amplitude and its dependence on pulse shape. (**a**) Normalized amplitude against roll-off factor (ROF) for raised-cosine (RC; blue) and square-root-raised-cosine (SRRC; red) signals. Simulations for maximum amplitude at eye-opening (solid lines), maximum amplitude between eye-openings (dashed line), and electrical measurements of the root-mean-square (RMS) amplitude (square markers). (**b**) Simulated eye-diagrams of RC and SRRC signal with roll-off factors of 0.1, 0.4, and 0.9. The color indicates the probability of occurrence; a logarithmic color mapping was used to enhance the visibility of small details; all plots are normalized to the same maximum amplitude d_DAC (digital-to-analog converter).

We analyzed the amplitude of the eye-opening for RC and SRRC and found different dependencies for the ROF. Figure 1 shows the results of this analysis, where we used a pulse-amplitude modulation (PAM) signal with two levels and determined the amplitude of the waveform. In Figure 1a, the amplitude over ROF is plotted for signals with RC pulse shape in blue and for signals with SRRC pulse shape in red. Here, we show the amplitude at the eye-opening as a solid line and the amplitude between the eye-openings as a dashed line. For the simulations, we always determined the maximum value at these positions, as this defines the scaling for the DAC. In Figure 1a, we also show experimentally determined results for the RMS amplitude plotted with square markers. The experimental results were linearly scaled to fit to the simulation.

For RC signals, the amplitude of the eye-opening increases with the ROF. This value is marked as d_{Eye} in Figure 1. The increase can be seen in simulations and measurements in blue in Figure 1a, as well as in the eye-diagrams for RC signals in Figure 1b. The reason for this is the decreasing swing in between the eye-openings. Since the waveform is normalized to this maximum value, this leads to an effective increase of the eye-opening. The dependence of the overshoot between two symbols for Nyquist filtering and the resulting restrictions were also discussed in Reference [31].

For SRRC signals, the behavior is different. When increasing the ROF, the amplitude of the eye-opening increases until 0.4, but does not further increase above that amplitude [54] (red square markers). After an ROF of slightly above 0.4, the maximum value between the eye-openings becomes smaller than the value at the eye-opening (intersection of red dashed and red solid line). For simplicity, we used 0.4 as the ROF for the intersection.

To illustrate the above-discussed behaviors, we show the impact to the signal generated by a DAC; Figure 1b shows simulated eye-diagrams for RC and SRRC signals with ROFs of 0.1, 0.4, and 0.9. All eye-diagrams were scaled to the same size, which depict the behavior the DAC would have while generating such a signal. The eye-opening for the RC signals increases from left to right, whereas, for SRRC signals, the eye-opening increases from an ROF of 0.1 to 0.4; however, at 0.4, the maximum value of the waveform is already at the eye-opening. Thus, there is no further increase of the RMS amplitude for an ROF larger than 0.4.

In conclusion, maximizing the OSNR in driver-amplifier-less transceivers can, in a first step, be realized by maximizing the electrical driving amplitude. For this, it was expected that increasing the ROF would be beneficial. However, for the case of SRRC signals, which are generally used for the transmitter, there is no benefit of increasing the ROF above 0.4. The analysis was done for a symbol rate which was well below the bandwidth limitations of the DAC. The simulations were done without restrictions of DAC resolution, effective number of bits (ENOB), SNR, and bandwidth. This is reasonable because, for this step, the influence of the pulse shape to the waveform was of interest and not the whole transceiver performance.

3. Investigation of the Effect of Clipping on the Electrical Eye-Opening

A reason for the roll-off factor dependence of the eye-opening as discussed in the previous section is the peak-to-average power ratio (PAPR). The different PAPRs can be seen in the simulated eye diagrams of Figure 1b. Especially, in the context of multicarrier systems [22], e.g., orthogonal frequency-division multiplexing (OFDM) [23,24,26,55], high PAPR is a well-discussed fundamental issue. Clipping is a solution to handle waveforms with a large PAPR [21–27]. Clipping limits the maximum amplitude to a certain level. This can be either achieved using a nonlinear driver amplifier or by clipping the values in the digital domain [22]. Clipping reduces the dynamical range, and thus, allows the reduction of the number of required bits for the analog-to-digital conversion [21]. Clipping, however, induces additional noise [21], and therefore, needs to be applied wisely.

Figure 2 shows a simulation for the impact of clipping on SRRC signals. We generated waveforms with one million PAM-2 symbols, oversampled with two samples per symbol, and SRRC as a pulse shape. The waveforms were clipped at different amplitude levels and received with a matched filter; finally, the SNR penalty was evaluated. Eight different clipping values between 0% and 33% clipping were chosen. These clipping values were chosen because they allowed an analysis of up to an SNR penalty of slightly more than 3 dB. A value of 0% clipping means no clipping at all, whereas 33% clipping means that the amplitude was saturated for values above 67% of the positive and negative maximal amplitude. Figure 2a shows the root-mean square (RMS) value of the signal amplitude normalized to the maximum value as a function of different roll-off factors. For a wide range, the RMS amplitude increases with clipping and roll-off factor. Figure 2b shows the resulting SNR penalty from clipping. The SNR penalty increases significantly with the roll-off factor. This is contrary to the optimum in RMS amplitude. Therefore, the ideal ROF lies in the area of 0.4, which allows a large RMS amplitude, but limits the SNR penalty due to clipping. The black crosses in Figure 2 indicate examples at ROFs of 0.2 and 0.4. The power for an ROF of 0.2 with 33% clipping is similar to the power for an ROF of 0.4 with clipping of only 14%. However, the SNR penalty at 0.2 with 33% of clipping is 1 dB, and for 0.4 with 14% clipping, the penalty is only 0.27 dB. This leads to the conclusion that clipping can help increase the RMS amplitude, and thus, the modulation efficiency of the EO conversion. However, a smart choice of ROF does further increase the signal quality. The ROF should not be blindly chosen and just compensated for by applying clipping.

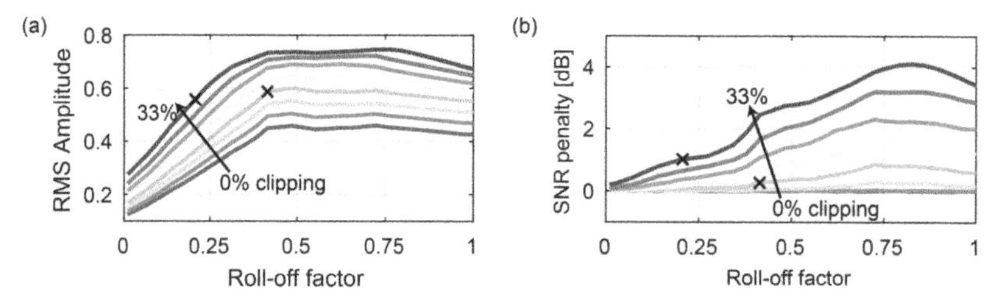

Figure 2. Investigation of clipping on signals with SRRC pulse shape. The simulations were done with two samples per symbol, one million symbols, 30 roll-off factors, and eight clipping values (0, 0.05, 0.10, 0.14, 0.19, 0.24, 0.29, and 0.33). The black crosses indicate examples at ROFs of 0.2 and 0.4, which are discussed in the text in more detail. (**a**) RMS amplitude of clipped and normalized (to maximum) waveforms against roll-off factor. Clipping was done between 0% up to 33%, which corresponds to a saturation of the amplitude at 100% down to 67%. (**b**) Resulting signal-to-noise ratio (SNR) penalty of clipped SRRC waveforms. The SNR was evaluated after matched filtered detection.

4. Influence of DPreD on the Electrical Eye-Opening

In this section, we describe how the complex frequency response of the link is measured, and we show the influence of DPreD on the RMS amplitude and the bit error ratio (BER) performance as derived from optical measurements. When speaking of DPreD, we correct the transmitter frequency response in the transmitter digital signal processing (DSP) before the digital-to-analog conversion; in the case of DPostD, the finite impulse response (FIR) filter correcting the transmitter frequency response is applied right at the beginning of the receiver DSP.

To measure the frequency and phase response of the link, we generated a test signal and measured its frequency and phase response at the receiver. The test signal consisted of a frequency comb with predefined phase relations. The best results were obtained for a frequency comb that was only constructed with spectral coefficients that correspond to prime numbers [56]. If equidistant spacing was taken, mixing products, e.g., due to non-linearities (NL), would have interfered with the desired spectral lines and possibly could have impaired the measurements. The phase relations of the spectral lines were chosen randomly; this reduced the peak-to-average power ratio of the test signal [57]. This signal was generated with the Micram DAC4 and measured in an electrical back-to-back configuration with an oscilloscope. The measured signal was Fourier-transformed and the difference to the transmitted signal was evaluated. The FIR filter impulse response for the DPreD, as well as that for the DPostD, was then determined by the inverse Fourier transform of the relative change in the coefficients.

Our experimental evaluations showed that the frequency comb is a good test signal to analyze the operation conditions of the set-up (e.g., for the bias voltage of the Mach–Zehnder modulator), minimizing the mixing products between the lines and maximizing the spectral lines lead to a linear operation point with the smallest non-linear mixing products.

Figure 3a shows the amplitude spectrum of a measured frequency comb with 34 spectral lines (chosen as discussed above) for the digitally designed signal in blue and the measured signal in red. The small number of spectral lines was chosen for depictive reasons only. It was previously shown that more spectral lines give an improvement in final signal quality after application of the correction filter. The frequency and phase response, as plotted in Figure 3b, were determined as the difference of the transmitted and received signal for a frequency comb with 331 spectral lines. This measurement shows 3-dB and 6-dB bandwidths of 33 GHz and 44 GHz, respectively.

Applying the correction filter to a new iteration of a comb-based frequency response measurement led to a flat response in frequency and phase. Afterward, the correction filter could be updated to slight changes in the experimental set-up or drifts by the impulse response of the least-mean-square equalizer of transmitted data. The update was done by a convolution of both impulse responses.

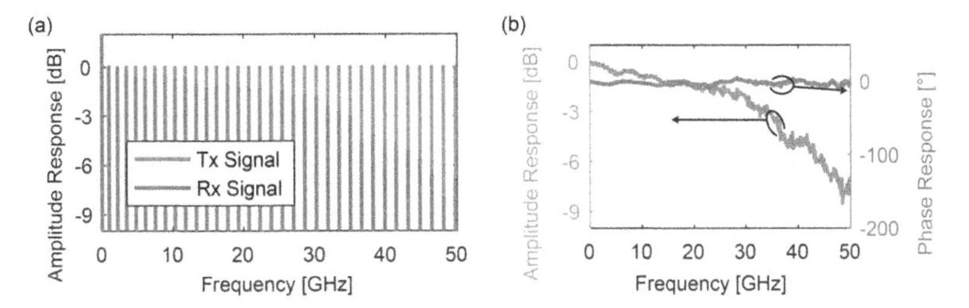

Figure 3. Characterization of frequency and phase response. (**a**) Amplitude spectrum of transmitted and received coarse non-integer spaced comb with a reduced set of 34 spectral lines. (**b**) Frequency and phase response measured with a comb of 331 spectral lines.

In the next step, we evaluated the influence of DPreD on the output of our DAC. The measurements were done with an electrical back-to-back set-up, where the RMS amplitude of the DAC output was directly measured with an oscilloscope with 160 GSa/s. Figure 4 shows a measurement series for SRRC signals with a range of different symbol rates and roll-off factors with and without DPreD. The colors indicate different symbol rates. For each symbol rate, roll-off factors between zero to one were tested. For symbol rates beyond 50 GBd, we swept the roll-off factor from zero until the highest possible, limited by $f_s/2 \geq B = 0.5(1 + ROF)R$, where f_s is the sampling frequency.

Three effects can be seen in the results. Firstly, the saturation for SRRC signals at a roll-off factor above 0.4, as described in Section 2, can be seen. Secondly, the influence of the DAC's frequency response to the signals without DPreD can be seen in Figure 4a. That is why the measured RMS amplitude decreases for higher signal bandwidths. The slope of the amplitude decrease follows the evaluated response from Figure 3b. Thirdly, the difference between Figure 4a,b shows the influence of DPreD on the electrical eye-opening. In particular, for high symbol rates, there is a significant attenuation of the electrical eye-opening by DPreD upon comparing it to the case where no DPreD was used.

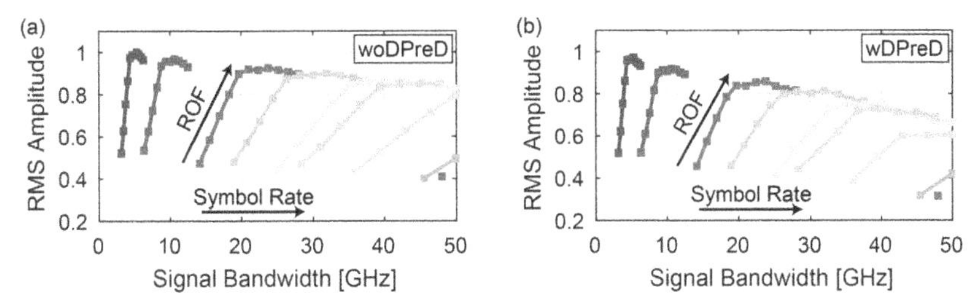

Figure 4. Electrical characterization of the influence of digital pre-distortion (DPreD) on the RMS amplitude of an SRRC signal generated with 100 GSa/s (**a**) without DPreD and (**b**) with DPreD. Both plots were linearly normalized to the same maximum value. The colors indicate the symbol rate (6.25, 12.5, 28, 37.5, 50, 56, 71, 80, 90, and 95 GBd). The signal bandwidth results from the symbol rate and the roll-off factor. woDPreD: without digital pre-distortion; wDPreD: with digital pre-distortion.

5. Measurements

We evaluated the influence of DPreD in a coherent optical link. We show 16QAM 28- and 56-GBd signals generated using the SRRC pulse shape. The signals were generated with a Micram DAC4 [58] with a sampling rate of 100 GSa/s, a 3-dB bandwidth of 35 GHz, and a frequency-dependent ENOB between 4 and 6 bits. The DAC was connected without driver amplifiers to an $LiNbO_3$ Oclaro modulator (V_π of 3.5 V, 3-dB bandwidth of 38 GHz). The measurements were done with a coherent receiver with a sampling rate of 160 GSa/s and a 3-dB bandwidth of 63 GHz. The performance was evaluated for different DAC output swings, roll-off factors, and symbol rates.

Figure 5 summarizes the applied DSP steps and gives an overview of the experimental set-up. In the transmitter (Tx) DSP, binary information was mapped to complex symbols, the desired pulse shape was applied, and, if desired, the DPreD was used to correct the frequency response. The experimental set-up consisted of the Micram DAC4 connected without electrical driver amplifiers to the IQ modulator, which modulated a laser operated at a wavelength of 1550 nm. The optical output of the modulator was amplified and filtered with a 2-nm bandpass filter before it was detected with a coherent receiver. In the receiver (Rx) DSP, DPostD was applied in the first step—if desired. Afterward, timing recovery [59], carrier recovery, least-mean-square equalization, hard decision demapping, and bit error ratio testing (BERT) were applied. For the analysis below, we report the BER we acquired using the BERT and the effective SNR, which we determined by comparing the transmitted symbols with the received symbols after the equalizer. The effective SNR allows modulation format-independent analyses and was, therefore, chosen for the final evaluations.

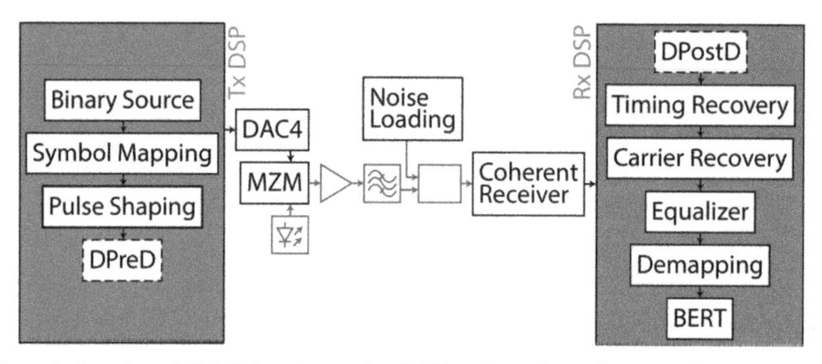

Figure 5. Overview of digital signal processing (DSP) and experimental set-up. The optical path is indicated by the blue color. Tx: transmitter; MZM: Mach–Zehnder modulator; Rx: receiver; BERT: bit error ratio testing.

Figure 6a,b show the influence of the DAC output voltage on the bit error ratio (BER) and the effective SNR of a 28-GBd 16QAM signal. The BER was determined from the ratio of wrongly transmitted bits to the total number of transmitted bits. The DAC output swing could be chosen between 350 and 750 mV_{pp}. We evaluated the performance in the low- and the high-SNR regime and show the results for DPreD in blue and DPostD in red. The SNR was adapted to the low- and high-SNR regime by optical noise loading. Due to the higher electrical signal power of signals with DPostD, there is an improvement in the low-SNR, as well as the high-SNR, regime when compared to DPreD. In both SNR regimes, there is an increase of 0.5 dB in effective SNR when changing from DPreD to DPostD. This corresponds to an increase of 12% in optical signal power.

Figure 6c shows the BER for a 56-GBd 16QAM signal measured for different roll-off factors. The performance increases with the roll-off factor. The slope of the improvement changes at an ROF of 0.4, which can be seen due to the given trend lines that are given below and above 0.4. The change in BER slope at a roll-off factor of 0.4 can be attributed to the amplitude of SRRC signals, as discussed in Section 2. We attribute the continued BER improvement above 0.4 to a wider horizontal eye-opening, which gives resilience toward jitter, and thus, a better BER.

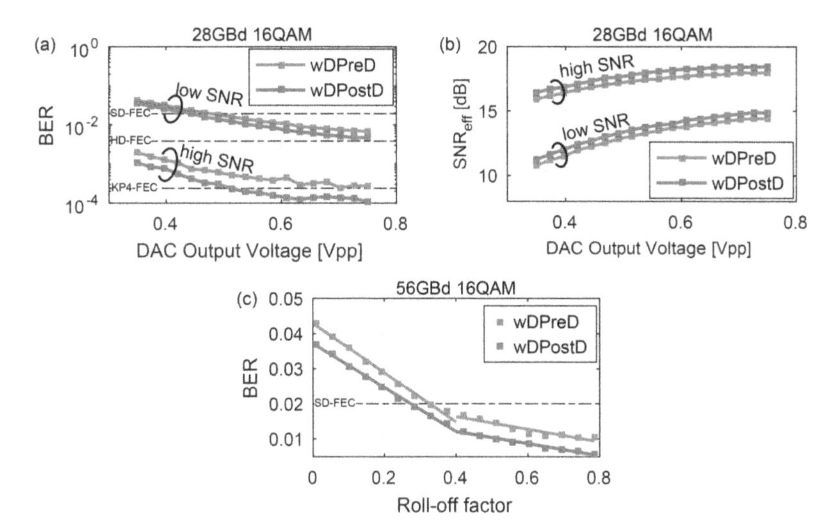

Figure 6. Influence of DPreD and digital post-distortion (DPostD) on the performance of an optical coherent 16QAM signal. Forward error correction (FEC) limits for KP4-FEC (BER: 2×10^{-4}, hard-decision FEC (HD-FEC) (BER: 3.8×10^{-3}), and soft-decision FEC (SD-FEC) (BER: 2×10^{-2}) are given. (**a**) BER over DAC driving voltage of a 28-GBd signal measured in the low- and high-SNR regime. (**b**) SNR over DAC driving voltage for the measurements in (**a**). (**c**) BER over roll-off factor of a 56-GBd signal. The solid line is a linear interpolation of the measurement values below and above a roll-off factor of 0.4. BER: bit error ratio; DAC: digital-to-analog converter.

To show the gain in overall signal quality due to DPostD, we measured, in a final step, the effective SNR for different symbol rates and ROFs, and compared the results of DPreD with the results of DPostD. Figure 7a shows the effective SNR for symbol rates of 14, 28, 56, 71, 80, and 96 GBd with roll-off factors of 0.04, 0.4, and 0.7 with DPreD as the compensation scheme. The possible symbol rate and roll-off factors are limited by the DAC sampling rate by $f_s/2 \geq B = 0.5(1 + ROF)R$, which explains the fewer measurement points for the ROFs of 0.4 and 0.7. Figure 7b shows the measurement results for compensation using DPostD. The trend follows the expected decrease of SNR for higher symbol rates. When increasing the symbol rate from 28 GBd to 56 GBd, there is a decrease in SNR of 3 dB, which is expected due to the fixed signal power, fixed noise spectral density, and the doubling of the signal bandwidth. In addition, it shows that the signal quality increases with ROF. However, when changing from 0.4 to 0.7, the increase is negligible, which follows the results of Section 2. Figure 7c shows the comparison of the measured effective SNR for DPreD to DPostD with a roll-off factor of 0.04. Over all measured symbol rates, there is an SNR gain of at least 0.5 dB, which shows the advantage of DPostD over DPreD for a wide range of symbol rates.

The results in this section clearly indicate that DPostD is beneficial for transmission links that use Mach-Zehnder modulators, which are operated with rather small electrical swings (such as in driver-amplifier-less systems). However, if the link is strongly influenced by nonlinear effects, linear DPostD might experience a penalty compared to DPreD. This is, e.g., the case for a driver amplifier limited by saturation, a modulator operated close to V_π, or a nonlinear fiber link. In these cases, the linear DPostD needs to be supported by a nonlinear compensation.

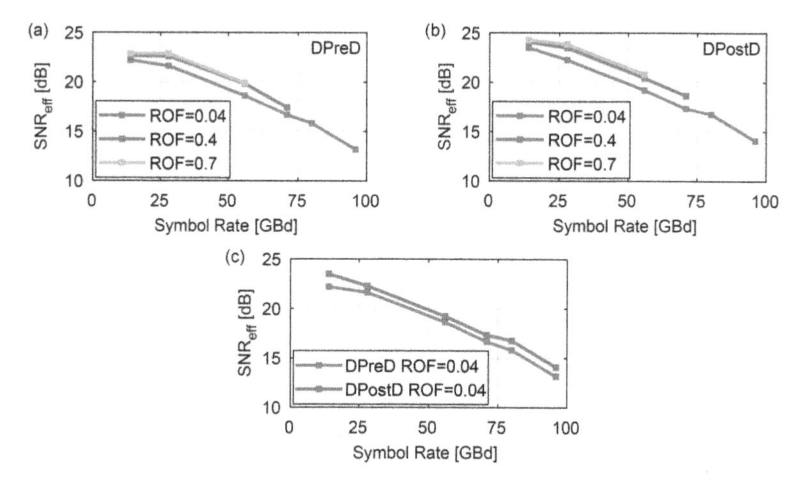

Figure 7. SNR improvement due to DPostD. Dependence of effective SNR on the symbol rate and ROF with DPreD (**a**) and with DPostD (**b**). (**c**) Comparison of effective SNR with an ROF of 0.04 with DPreD in blue and DPostD in red. The measured symbol rates are 14, 28, 56, 71, 80, and 96 GBd; QPSK was used as the modulation format.

6. Conclusions

We investigated the prerequisites for cost-efficient driver-amplifier-less optical transmitters. Without driver amplifiers, a signal with maximum electrical eye-opening is advantageous. For this, we investigated the influence of the pulse shape, clipping, and DPreD on the electrical eye-opening. We showed that increasing the ROF of SRRC signals up to an ROF of 0.4 is advantageous, but there is little advantage for ROFs beyond 0.4. We also showed that DPostD results in better BER performances than DPreD.

Author Contributions: Conceptualization, A.J., B.B., and J.L. Data curation, A.J. and W.H. Formal analysis, A.J., B.B., and J.L. Investigation, A.J., B.B., R.B., W.H., and J.L. Methodology, A.J., B.B., R.B., and J.L. Project administration, A.J. and J.L. Resources, J.L. Software, A.J. and B.B. Supervision, J.L. Validation, A.J., B.B., R.B., and J.L. Visualization, A.J. and R.B. Writing—original draft, A.J. Writing—review and editing, A.J., B.B., R.B., W.H., and J.L.

Funding: The EU-project PLASMOfab (688166) and the ERC PLASILOR (670478) are acknowledged for financial support.

Acknowledgments: We thank Oclaro for supplying us with the DP IQ modulators. We also thank unknown reviewers that helped improve the discussions of this manuscript with their valuable feedback.

Conflicts of Interest: The authors declare no conflicts of interest.

References

1. Miller, D.A.B. Attojoule optoelectronics for low-energy information processing and communications. *J. Lightw. Technol.* **2017**, *35*, 346–396. [CrossRef]
2. Fujisawa, T.; Kanazawa, S.; Takahata, K.; Kobayashi, W.; Tadokoro, T.; Ishii, H.; Kano, F. 13-μm, 4 × 25-gbit/s, eadfb laser array module with large-output-power and low-driving-voltage for energy-efficient 100 gbe transmitter. *Opt. Express* **2011**, *20*, 614. [CrossRef] [PubMed]
3. Kataoka, T.; Miyamoto, Y.; Hagimoto, K.; Wakita, K.; Kotaka, I. Ultrahigh-speed driverless mqw intensity modulator, and 20 gbit/s, 100 km transmission experiments. *Electron. Lett.* **1992**, *28*, 897–898. [CrossRef]
4. Loi, K.K.; Shen, L.; Wieder, H.H.; Chang, W.S.C. Novel high-frequency electroabsorption multiple-quantum-well waveguide modulator operating at 1.3 μm on gaas substrates. *Optoelectron. Integr. Circ.* **1997**. [CrossRef]

5. Wang, L.; Hu, R.; Li, M.; Qiu, Y.; Chen, D.; Xiao, X.; Li, Z.; Yu, Y.; Yu, J.; Yang, Q.; et al. Transmission of 24-Gb/s PAM-4 over 150-km SSMF using a driverless silicon microring modulator. In Proceedings of the Asia Communications and Photonics Conference 2014, Shanghai, China, 11–14 November 2014.
6. Ackerman, E.; Betts, G.; Burns, W.; Prince, J.; Regan, M.; Roussell, H.; Cox, C. Low noise figure, wide bandwidth analog optical link. In Proceedings of the 2005 International Topical Meeting on Microwave Photonics, Seoul, Korea, 14 October 2005; pp. 325–328.
7. Cox, C.H.; Ackerman, E.I.; Betts, G.E.; Prince, J.L. Limits on the performance of RF-over-fiber links and their impact on device design. *IEEE Trans. Microw. Theory Tech.* **2006**, *54*, 906–920. [CrossRef]
8. Zhou, W.; Okusaga, O.; Nelson, C.; Howe, D.; Carter, G. 10 ghz dual loop opto-electronic oscillator without rf-amplifiers. In Proceedings of the Optoelectronic Devices, San Jose, CA, USA, 19–24 January 2008. [CrossRef]
9. Shastri, A.; Webster, M.; Jeans, G.; Metz, P.; Sunder, S.; Chattin, B.; Dama, B.; Shastri, K. Experimental demonstration of ultra-low-power single polarization 56 Gb/s QAM-16 generation without DAC using CMOS photonics. In Proceedings of the 2014 The European Conference on Optical Communication (ECOC), Cannes, France, 21–25 September 2014; pp. 1–3.
10. Wolf, S.; Lauermann, M.; Schindler, P.; Ronniger, G.; Geistert, K.; Palmer, R.; Kober, S.; Bogaerts, W.; Leuthold, J.; Freude, W.; Koos, C. Dac-less amplifier-less generation and transmission of qam signals using sub-volt silicon-organic hybrid modulators. *J. Light. Technol.* **2015**, *33*, 1425–1432. [CrossRef]
11. Fujisawa, T.; Kanazawa, S.; Takahata, K.; Kobayashi, W.; Tadokoro, T.; Ishii, H.; Kano, F. Large-Output-Power, Ultralow-Driving-Voltage (0.5 Vpp) Operation of 1.3-um, 4 × 25 G, EADFB laser array for driverless 100 Gbe transmitter. In Proceedings of the European Conference on Optical Communication 2011, Geneva, Switzerland, 18–22 September 2011.
12. Palmer, R.; Alloatti, L.; Korn, D.; Schindler, P.C.; Baier, M.; Bolten, J.; Wahlbrink, T.; Waldow, M.; Dinu, R.; Freude, W.; et al. Low power mach-zehnder modulator in silicon-organic hybrid technology. *IEEE Photonics Technol. Lett.* **2013**, *25*, 1226–1229. [CrossRef]
13. Timurdogan, E.; Sorace-Agaskar, C.M.; Sun, J.; Hosseini, E.S.; Biberman, A.; Watts, M.R. An ultralow power athermal silicon modulator. *Nat. Commun.* **2014**, *5*, 4008. [CrossRef] [PubMed]
14. Koeber, S.; Palmer, R.; Lauermann, M.; Heni, W.; Elder, D.L.; Korn, D.; Woessner, M.; Alloatti, L.; Koenig, S.; Schindler, P.C.; et al. Femtojoule electro-optic modulation using a silicon–organic hybrid device. *Light. Sci. Appl.* **2015**, *4*, e255. [CrossRef]
15. Nozaki, K.; Shakoor, A.; Matsuo, S.; Fuji, T.; Takeda, K.; Shinya, A.; Kuramochi, A.; Notomi, M. Ultralow-energy electro-absorption modulator consisting of InGaAsP-embedded photonic-crystal waveguide. *APL Photonics* **2017**, *2*, 056105. [CrossRef]
16. Baeuerle, B.; Hoessbacher, C.; Heni, W.; Fedoryshyn, Y.; Josten, A.; Haffner, C.; Watanabe, T.; Elder, D.L.; Dalton, L.R.; Leuthold, J. Driver-less Sub-1 Vpp-operation of a Plasmonic-organic hybrid modulator at 100 GBd NRZ. In Proceedings of the 2018 Optical Fiber Communications Conference and Exposition (OFC), San Diego, CA, USA, 11–15 March 2018.
17. Sowailem, M.Y.S.; Hoang, T.M.; Morsy-Osman, M.; Chagnon, M.; Patel, D.; Paquet, S.; Paquet, C.; Woods, I.; Liboiron-Ladouceur, O.; Plant, D.V. 400-G single carrier 500-km transmission with an InP dual polarization IQ modulator. *IEEE Photonics Technol. Lett.* **2016**, *28*, 1213–1216. [CrossRef]
18. Bülow, H.; Buchali, F.; Klekamp, A. Electronic dispersion compensation. *J. Lightw. Technol.* **2008**, *26*, 158–167. [CrossRef]
19. Sowailem, M.Y.S.; Hoang, T.M.; Morsy-Osman, M.; Chagnon, M.; Qiu, M.; Paquet, S.; Paquet, C.; Woods, I.; Liboiron-Ladouceur, O.; Plant, D.V. Impact of chromatic dispersion compensation in single carrier two-dimensional stokes vector direct detection system. *IEEE Photonics J.* **2017**, *9*, 1–10. [CrossRef]
20. Zhou, X.; Huo, J.; Zhong, K.; Khan, F.N.; Gui, T.; Zhang, H.; Tu, J.; Yuan, J.; Long, K.; Yu, C.; et al. Single channel 50 Gbit/s transmission over 40 km SSMF without optical amplification and in-line dispersion compensation using a single-end PD-based PDM-SSB-DMT system. *IEEE Photonics J.* **2017**, *9*, 1–11. [CrossRef]
21. Mestdagh, D.J.G.; Spruyt, P.; Biran, B. Analysis of clipping effect in DMT-based ADSL systems. In Proceedings of the 1994 International Conference on Communications, ICC/SUPERCOMM/ICC '94, Conference Record, Serving Humanity Through Communications, New Orleans, LA, USA, 1–5 May 1994.
22. Tellado, J. *Mutlicarrier modulation with low PAR: Applications to DSL and Wireless*, 1st ed.; Kluwer Academic Publishers: Dordrecht, The Netherlands, 2002; ISBN 0-7923-7988-8.

23. Armstrong, J. OFDM for Optical Communications. *J. Lightw. Technol.* **2009**, *27*, 189–204. [CrossRef]
24. Hwang, T.; Yang, C.Y.; Wu, G.; Li, S.; Li, G.Y. OFDM and its wireless applications: A survey. *IEEE Trans. Veh. Technol.* **2009**, *58*, 1673–1694. [CrossRef]
25. Lim, D.W.; Heo, S.J.; No, J.S. An Overview of peak-to-average power ratio reduction schemes for OFDM signals. *J. Commun. Netw.* **2009**, *11*, 229–239. [CrossRef]
26. Shieh, W.; Djordjevic, I. *OFDM for Optical Communications*; Academic Press: Burlington, MA, USA, 2009.
27. Randel, S.; Breyer, F.; Lee, S.C.J.; Walewski, J.W. Advanced modulation schemes for short-range optical communications. *IEEE J. Sel. Top. Quantum Electron.* **2010**, *16*, 1280–1289. [CrossRef]
28. Fludger, C.R.S.; Duthel, T.; Hermann, P.; Kupfer, T. Low cost transmitter self-calibration of time delay and frequency response for high baud-rate QAM transceivers. In Proceedings of the Optical Fiber Communication Conference 2017, Los Angeles, CA, USA, 19–23 March 2017.
29. Fludger, C.R.; Vercelli, E.S.; Marenco, G.; Torre, A.D.; Duthel, T.; Kupfer, T. 1Tb/s real-time 4 × 40 Gbaud DP-16QAM superchannel using CFP2-ACO pluggables over 625 km of standard fibre. In Proceedings of the Optical Fiber Communication Conference 2016, Anaheim, CA, USA, 20–22 March 2016.
30. Juan, Q.; Mao, B.; Gonzalez, N.; Binh, N.; Stojanovic, N. Generation of 28 GBaud and 32 GBaud PDM-Nyquist-QPSK by a DAC with 11.3 GHz analog bandwidth. In Proceedings of the Optical Fiber Communication Conference 2013, Anaheim, CA, USA, 17–21 March 2013.
31. Buchali, F.; Klekamp, A.; Schmalen, L.; Drenski, T. Implementation of 64QAM at 42.66 GBaud using 1.5 samples per symbol DAC and demonstration of up to 300 km fiber transmission. In Proceedings of the Optical Fiber Communication Conference 2014, San Francisco, CA, USA, 9–13 March 2014.
32. Rafique, D.; Napoli, A.; Calabro, S.; Spinnler, B. Digital preemphasis in optical communication systems: On the dac requirements for terabit transmission applications. *J. Light. Technol.* **2014**, *32*, 3247–3256. [CrossRef]
33. Chagnon, M.; Morsy-Osman, M.; Poulin, M.; Paquet, C.; Lessard, S.; Plant, D.V. Experimental parametric study of a silicon photonic modulator enabled 112-gb/s pam transmission system with a dac and adc. *J. Light. Technol.* **2015**, *33*, 1380–1387. [CrossRef]
34. Napoli, A.; Mezghanni, M.M.; Rahman, T.; Rafique, D.; Palmer, R.; Spinnler, B.; Calabrò, S.; Castro, C.; Kuschnerov, M.; Bohn, M. Digital compensation of bandwidth limitations for high-speed DACs and ADCs. *J. Lightw. Technol.* **2016**, *34*, 3053–3064. [CrossRef]
35. Khanna, G.; Spinnler, B.; Calabro, S.; De Man, E.; Hanik, N. A robust adaptive pre-distortion method for optical communication transmitters. *IEEE Photonics Technol. Lett.* **2016**, *28*, 752–755. [CrossRef]
36. Khanna, G.; Rahman, T.; Man, E.D.; Riccardi, E.; Pagano, A.; Piat, A.C.; Calabro, S.; Spinnler, B.; Rafique, D.; Feiste, U.; et al. Single-carrier 400 g 64qam and 128qam dwdm field trial transmission over metro legacy links. *IEEE Photonics Technol. Lett.* **2017**, *29*, 189–192. [CrossRef]
37. Khanna, G.; Spinnler, B.; Calabro, S.; de Man, E.; Chen, Y.; Hanik, N. Adaptive transmitter pre-distortion using feedback from the far-end receiver. *IEEE Photonics Technol. Let.* **2018**, *30*, 223–226. [CrossRef]
38. Changsoo, E.; Powers, E.J. A new volterra predistorter based on the indirect learning architecture. *IEEE Trans. Signal Process.* **1997**, *45*, 223–227. [CrossRef]
39. Guiomar, F.P.; Pinto, A.N. Simplified volterra series nonlinear equalizer for polarization-multiplexed coherent optical systems. *J. Lightw. Technol.* **2013**, *31*, 3879–3891. [CrossRef]
40. Berenguer, P.W.; Nolle, M.; Molle, L.; Raman, T.; Napoli, A.; Schubert, C.; Fischer, J.K. Nonlinear digital pre-distortion of transmitter components. *J. Lightw. Technol.* **2016**, *34*, 1739–1745. [CrossRef]
41. Jia, Z.; Chien, H.-C.; Cai, Y.; Yu, J.; Zhu, B.; Ge, C.; Wang, T.; Shi, S.; Wang, H.; Xia, Y.; et al. Experimental demonstration of PDM-32QAM single-carrier 400G over 1200-km transmission enabled by training-assisted pre-equalization and look-up table. In Proceedings of the 2016 Optical Fiber Communication Conference (OFC), Anaheim, CA, USA, 20–24 March 2016.
42. Junwen, Z.; Yu, J.; Chien, H.-C. Advanced Algorithm for high-baud rate signal generation and detection. In Proceedings of the Optical Fiber Communication Conference 2017, Los Angeles, CA, USA, 19–23 March 2017.
43. Napoli, A.; Mezghanni, M.M.; Calabrò, S.; Palmer, R.; Saathoff, G.; Spinnler, B. Digital predistortion techniques for finite extinction ratio IQ mach–zehnder modulators. *J. Lightw. Technol.* **2017**, *35*, 4289–4296. [CrossRef]
44. Yamazaki, H.; Takahashi, H.; Goh, T.; Hashizume, Y.; Yamada, T.; Mino, S.; Kawakami, H.; Miyamoto, Y. Optical modulator with a near-linear field response. *J. Lightw. Technol.* **2016**, *34*, 3796–3802. [CrossRef]

45. Josten, A.; Baeuerle, B.; Heni, W.; Leuthold, J. Digital post-distortion for cost-efficient driverless optical transmitters. In Proceedings of the Signal Processing in Photonic Communications 2018, Zurich, Switzerland, 2–5 July 2018.
46. Gnauck, A.H.; Raybon, G.; Chandrasekhar, S.; Leuthold, J.; Doerr, C.; Stulz, L.; Agarwal, A.; Banerjee, S.; Grosz, D.; Hunsche, S.; et al. 2.5 Tb/s (64 × 42.7 Gb/s) Transmission over 40 × 100 km NZDSF using RZ-DPSK formatand all-raman-amplified spans. In Proceedings of the Optical Fiber Communications Conference, Anaheim, CA, USA, 17 March 2002.
47. Winzer, P.J.; Dorrer, C.; Essiambre, R.-J.; Kang, I. Chirped return-to-zero modulation by imbalanced pulse carver driving signals. *IEEE Photonics Technol. Lett.* **2004**, *16*, 1379–1381. [CrossRef]
48. Gnauck, A.H.; Winzer, P.J. Optical Phase-Shift-Keyed Transmission. *J. Lightw. Technol.* **2005**, *23*, 115.
49. Kang, I.; Mollenauer, L.F. Method and Apparatus for Synchronizing a Pulse Carver and a Data Modulator for Optical Telecommunication. U.S. Patent 7,209,669 B2, 24 April 2007.
50. Behrens, C.; Makovejs, S.; Killey, R.I.; Savory, S.J.; Chen, M.; Bayvel, P. Pulse-shaping versus digital backpropagation in 224gbit/s pdm-16qam transmission. *Opt. Express* **2011**, *19*, 12879. [CrossRef] [PubMed]
51. Hirooka, T.; Ruan, P.; Guan, P.; Nakazawa, M. Highly dispersion-tolerant 160 Gbaud optical nyquist pulse tdm transmission over 525 km. *Opt. Express* **2012**, *20*, 15001. [CrossRef] [PubMed]
52. Soto, M.A.; Alem, M.; Amin Shoaie, M.; Vedadi, A.; Brès, C.-S.; Thévenaz, L.; Schneider, T. Optical sinc-shaped Nyquist pulses of exceptional quality. *Nat. Commun.* **2013**, *4*, 2898. [CrossRef] [PubMed]
53. Proakis, J.G.; Salehi, M. *Digital Communications*, 5th ed.; McGraw-Hill: New York, NY, USA, 2008.
54. Chatelain, B.; Gagnon, F. Peak-to-average power ratio and intersymbol interference reduction by Nyquist pulse optimization. In Proceedings of the IEEE 60th Vehicular Technology Conference, 2004, VTC2004-Fall, Los Angeles, CA, USA, 26–29 September 2004; pp. 954–958.
55. Jansen, S.L.; Morita, I.; Schenk, T.C.W.; Takeda, N.; Tanada, H. Coherent Optical 25.8-Gb/s OFDM transmission over 4160-km SSMF. *J. Lightw. Technol.* **2008**, *26*, 6–15. [CrossRef]
56. Buchali, F.; Idler, W.; Rastegardoost, N.; Drenski, T.; Ward, R.; Zhao, L. Preemphased prime frequency multicarrier bases ENOB assessment and its application for optimizing a dual-carrier 1-Tb/s QAM transmitter. In Proceedings of the 2016 Optical Fiber Communications Conference and Exhibition (OFC), Anaheim, CA, USA, 20–24 March 2016.
57. Nikookar, H.; Lidsheim, K.S. Random phase updating algorithm for ofdm transmission with low papr. *IEEE Trans. Broadcast.* **2002**, *48*, 123–128. [CrossRef]
58. Schuh, K.; Buchali, F.; Idler, W.; Hu, Q.; Templ, W.; Bielik, A.; Altenhain, L.; Langenhagen, H.; Rupeter, J.; Duemler, U.; et al. 100 GSa/s BiCMOS DAC supporting 400 Gb/s dual channel transmission. In Proceedings of the ECOC 2016; 42nd European Conference on Optical Communication, Dusseldorf, Germany, 18–22 September 2016.
59. Josten, A.; Baeuerle, B.; Dornbierer, E.; Boesser, J.; Hillerkuss, D.; Leuthold, J. Modified godard timing recovery for non integer oversampling receivers. *Appl. Sci.* **2017**, *7*, 655. [CrossRef]

Article

Performance Improvement for Mixed RF–FSO Communication System by Adopting Hybrid Subcarrier Intensity Modulation

Ting Jiang [1], Lin Zhao [1], Hongzhan Liu [1,2,*], Dongmei Deng [1,2], Aiping Luo [1,2], Zhongchao Wei [1,2] and Xiangbo Yang [1,2]

1 Guangdong Provincial Key Laboratory of Nanophotonic Functional Materials and Devices, Guangzhou 510006, China
2 School of Information and Optoelectronic Science and Engineering, South China Normal University, Guangzhou 510006, China
* Correspondence: lhzscnu@163.com

Received: 23 July 2019; Accepted: 4 September 2019; Published: 6 September 2019

Abstract: The improvement for hybrid radio frequency–free space optical (RF–FSO) communication system in wireless optical communications has acquired growing interests in recent years, but rarely improvement is based on hybrid modulation. Therefore, we conduct a research on end-to-end mixed RF–FSO system with the hybrid pulse position modulation–binary phase shift keying–subcarrier intensity modulation (PPM–BPSK–SIM) scheme. The RF link obeys Rayleigh distribution and the FSO link experiences Gamma–Gamma distribution. The average bit error rate (BER) for various PPM–BPSK–SIM schemes has been derived with consideration of atmospheric turbulence influence and pointing error condition. The outage probability and the average channel capacity of the system are discussed as well. Simulation results indicate that the pointing error aggravates the influence of atmospheric turbulence on the channel capacity, and the RF–FSO systematic performance is improved obviously while adopting PPM–BPSK–SIM under strong turbulence and severe pointing error conditions, especially, when the system average symbol length is greater than eight.

Keywords: free space optical (FSO); pulse position modulation–binary phase shift keying–subcarrier intensity modulation (PPM–BPSK–SIM); bit error rate (BER); pointing error; average symbol length

1. Introduction

In the past ten years, the hybrid radio frequency–free space optical (RF–FSO) communication system has attracted great interest due to its advantages compared with FSO system and RF system, respectively. RF system has the advantage of being insensitive to atmospheric turbulence, but the spectrum resource it requires is in shortage. On the contrary, features of the FSO system include license-free transmission, cost-effectiveness, and high security, which meet the demands of modern optical communication, but its communication quality is easily disturbed by the atmosphere [1,2]. The hybrid RF–FSO system could combine the advantages of the RF system and the FSO system [3]. Inevitably, dynamic weather and complex environment still affect the performance of hybrid RF–FSO system. Therefore, not only should a suitable channel model be chosen for hybrid RF–FSO system but also appropriate modulations could be adopted to improve the performance of the hybrid RF–FSO system.

Considering impacts of atmosphere including turbulence, pointing error, and pass fading, many channel models have been used to describe the hybrid RF–FSO system. Such as Nakagami-m/ Gamma–Gamma distribution [4], Rayleigh distribution and M-distributed distribution [5], $\kappa-\mu/\eta-\mu$ distribution and Gamma–Gamma distribution [6], and the Nakagami-m/Exponentiated Weibull (EW) distribution [7]. However, many studies adopted Rayleigh/Gamma–Gamma distribution [8–10], because the Gamma–Gamma distribution has a good simulation for the atmospheric turbulence from weak to strong of the FSO link, and Rayleigh distribution is close to the condition of the RF channel. As a result, we adopted Rayleigh distribution for the RF link and Gamma–Gamma distribution for the FSO link.

On the other hand, many previous researchers have focused on using different modulations to improve the mixed RF–FSO systematic performance for the modulation has a direct relationship with bit error rate (BER), which is a significant indicator to evaluate system performance. In reference [11,12], various pulse position modulation (PPM), binary phase shift keying (BPSK), and hybrid pulse position modulation–binary phase shift keying–subcarrier intensity modulation (PPM–BPSK–SIM) schemes have been analyzed on the performance over FSO link, the conclusion is that hybrid PPM–BPSK–SIM outperformed BPSK in single FSO link. In literature about the hybrid RF–FSO system [13,14], most of them adopted various binary modulation when it refers to the average bit error rate, BPSK is the most discussed as the binary modulation for it has the lowest bit error rate. In reference [4], authors investigated various PSK modulation and various quadrature amplitude modulation (QAM) in hybrid RF–FSO system. Unlike the FSO system [15], there are few studies on how to improve the hybrid RF–FSO system performance by adopting hybrid modulation, especially for the research of PPM–BPSK–SIM.

Motivated by the aforementioned studies, we investigated end-to-end performance of hybrid RF–FSO system with application of hybrid PPM–BPSK–SIM. The RF link follows Rayleigh distribution, the FSO link obeys Gamma–Gamma distribution, not only atmospheric turbulence but also pointing error is taken into consideration, and the system adopts various hybrid PPM–BPSK–SIM, the receiver employs intensity modulation/direct detection (IM/DD). The unconditional bit error rate (BER) of the PPM–BPSK–SIM is derived, it is noteworthy that the application of hybrid PPM–BPSK–SIM does bring optimization to the hybrid RF–FSO system performance from simulation results. The impact of pointing error on systematic performance is also discussed in detail.

The content of the article is arranged as follows. Specific information of the hybrid RF–FSO system and the channel model is described in Section 2. We derived the outage probability, the unconditional BER, and average channel capacity of the hybrid system in Section 3. The simulation is realized by MATLAB, and the results and discussion are shown in Section 4. Section 5 contains concluding remarks.

2. System and Channel Model

We consider an end-to-end hybrid RF–FSO system, it mainly consists of three parts—a source and a destination with a relay between them, the system model is shown below in Figure 1. The signal is received by the relay from the source through the RF channel, which obeys Rayleigh distribution. Considering the relatively lower implementation complexity [16], we adopt a fixed-gain amplify-and-forward (AF) relay which amplifies the RF signals and forwards them to next node directly [12]. In other words, the received RF signal will be transformed into an optical signal by the relay using the SIM technique, and the optical signal will be amplified by a fixed gain, G. After that, it is transmitted to the destination through the FSO link, which is supposed to obey Gamma–Gamma distribution, and IM/DD is used as the detection method in the destination.

The received signal of the relaying and the destination can be expressed $y_1 = h_1 x + n_1$ and $y_2 = h_2 y_1 + n_2$, so the comprehensive expression is $y_2 = h_2(h_1 x + n_1) + n_2$, where y_1 and y_2 denote the received signal, x represents the normalized signal from signal source, h_1 and h_2 are the channel coefficient of the RF link and the FSO link, respectively. $h_2 = IG\eta$, where I represents the irradiance intensity of FSO channel, G denotes the fixed gain of relaying scheme, and η represents the conversion coefficient of electrical-to-optical. n_1 and n_2 are additive white Gaussian noise (AWGN) with ${\sigma_1}^2$ and ${\sigma_2}^2$ as variance and mean of zero.

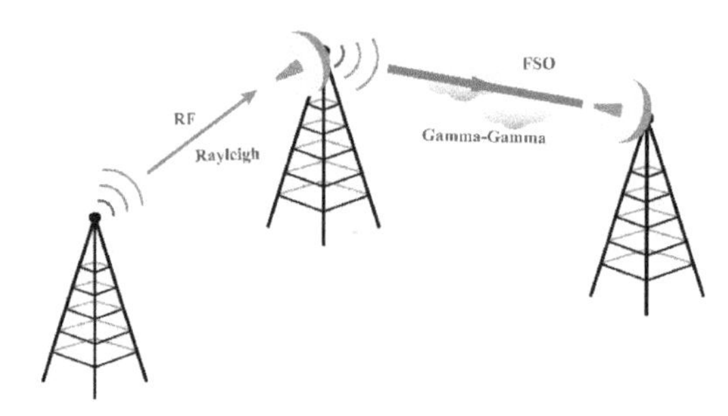

Figure 1. End-to-end mixed radio frequency (RF) and free space optical (FSO) communication system.

The RF link follows Rayleigh distribution in the hybrid RF–FSO system, the probability density function (PDF) of the RF link could be given as $f_{\gamma_1}(\gamma_1) = \frac{1}{\overline{\gamma_1}}\exp\left(-\frac{\gamma_1}{\overline{\gamma_1}}\right)$, where γ_1 represents the instantaneous signal to noise ratio (SNR) of the RF link and $\overline{\gamma_1}$ denotes the average SNR in the RF channel. Using the definition of the cumulative distribution function (CDF), $F_1 = \int_0^{\gamma_1} f_{\gamma_1}(\gamma)d\gamma_1$. Hence, the CDF of the Rayleigh distribution can be expressed as $F_1(\gamma_1) = 1 - \exp(\frac{\gamma_1}{\overline{\gamma_1}})$.

The FSO link obeys Gamma–Gamma distribution with consideration of pointing error and atmospheric turbulence. The PDF of the SNR in FSO link can be given by [14,17–22]:

$$f_{\gamma_2}(\gamma_2) = \frac{\xi^2}{2\gamma_2\Gamma(\alpha)\Gamma(\beta)} \times G\begin{matrix}3,0\\1,3\end{matrix}\left(\alpha\beta\sqrt{\frac{\gamma_2}{\overline{\gamma_2}}}\middle|\begin{matrix}\xi^2+1\\ \xi^2,\alpha,\beta\end{matrix}\right), \tag{1}$$

where γ_2 represents the instantaneous SNR of the FSO hop, $\overline{\gamma_2}$ is the average SNR in the FSO channel, α and β are the fading parameters that can reflect turbulence conditions. $\xi = \frac{\omega_e}{2\sigma_s}$, ξ is the pointing error parameter, which is determined by the pointing error displacement standard deviation at the destination and equivalent beam radius (i.e., when $\xi \to \infty$, we get a case without influence of pointing error, the pointing error can be ignored). $\Gamma(\cdot)$ is the Gamma function and $G(\cdot)$ is the Meijer's G function, and 3, 0, 1, 3 are the parameters associated with the Meijer's G function, they are defined in reference [23]. Using the same method as in RF link, the CDF of the Gamma–Gamma distribution with the influence of pointing error can be given by [24,25]:

$$\begin{aligned} F_2(\gamma_2) &= \int_0^{\gamma_2} f_{\gamma_2}(\gamma_2)d\gamma_2 \\ &= \frac{\xi^2}{2\gamma_2\Gamma(\alpha)\Gamma(\beta)} G\begin{matrix}3,1\\2,4\end{matrix}\left[Z\middle|\begin{matrix}2,\xi^2+1\\ \xi^2,\alpha,\beta,1\end{matrix}\right], \end{aligned} \tag{2}$$

where $Z = \frac{\alpha\beta\gamma_2}{\sqrt{\overline{\gamma_2}}}$.

Considering the relay with fixed gain, the signal will be amplified with a fixed relay gain, G, and forwarded to the destination. Hence, the end-to-end instantaneous SNR of the mixed RF–FSO system could be expressed as $\gamma = \frac{\gamma_1\gamma_2}{\gamma_2+C}$, where C is a constant associated with G [26].

The CDF of the hybrid RF–FSO system is defined as [9]:

$$\begin{aligned} F_\gamma(\gamma) &= \Pr\left(\frac{\gamma_1\gamma_2}{\gamma_2+C} < \gamma_{th}\right) \\ &= \int_0^\infty \Pr\left(\frac{\gamma_1\gamma_2}{\gamma_2+C} < \gamma_{th}\middle|\gamma_2\right)f_{\gamma_2}(\gamma_2)d\gamma_2 \\ &= \int_0^\infty F_1\left(\frac{\gamma_{th}(\gamma_2+C)}{\gamma_2}\right)f_{\gamma_2}(\gamma_2)d\gamma_2. \end{aligned} \tag{3}$$

Using $\Gamma(g,x) = \int_0^x t^{s-1}e^{-t}dt$, the CDF of Rayleigh distribution, and Equation (3), the CDF of the hybrid RF–FSO system is

$$F_\gamma(\gamma) = 1 - \frac{2^{(\alpha+\beta)}A\sqrt{C}}{16\pi\sqrt{\overline{\gamma_1}}}\sqrt{\gamma}\exp(-\frac{\gamma}{\overline{\gamma_1}}) \times G\,{}^{6,0}_{1,6}\left(\frac{(\alpha\beta)^2 C\gamma}{16\overline{\gamma_1}\overline{\gamma_2}}\middle|\begin{matrix}\frac{\xi^2+1}{2}\\ \kappa_1\end{matrix}\right), \tag{4}$$

where $A = \alpha\beta\xi^2/\left(2\sqrt{\overline{\gamma_2}}\Gamma(\alpha)\Gamma(\beta)\right)$ and $\kappa_1 = \frac{\xi^2-1}{2}, \frac{\alpha-1}{2}, \frac{\alpha}{2}, \frac{\beta-1}{2}, \frac{\beta}{2}, -\frac{1}{2}$.

3. Performance Analysis

3.1. Outage Probability

Outage probability is the probability that the instantaneous SNR, γ, is smaller than the threshold value γ_{th}, so the mathematical formula of it in the mixed RF–FSO system is defined as $P_{out}(\gamma_{th}) = \Pr(\gamma \le \gamma_{th})$.

Hence, the expression of outage probability can be obtained by $P_{out}(\gamma_{th}) = F_\gamma(\gamma_{th})$,

$$P_{out}(\gamma_{th}) = 1 - \frac{2^{(\alpha+\beta)}A\sqrt{C}}{16\pi\sqrt{\overline{\gamma_1}}}\sqrt{\gamma_{th}}\exp(-\frac{\gamma_{th}}{\overline{\gamma_1}}) \times G\,{}^{6,0}_{1,6}\left(\frac{(\alpha\beta)^2 C\gamma_{th}}{16\overline{\gamma_1}\overline{\gamma_2}}\middle|\begin{matrix}\frac{\xi^2+1}{2}\\ \kappa_1\end{matrix}\right). \tag{5}$$

3.2. Average Bit Error Rate

There are many modulations that can be used in the hybrid RF–FSO system, they have a large difference in systematic performance due to their different characteristics, such as power efficiency, bandwidth efficiency, and simplicity, the most intuitive way to evaluate the quality of a modulation method is the BER of the system.

The PPM–BPSK–SIM combines PPM and BPSK–SIM together, the symbol is modulated by a PPM encoder, then the parallel signal is transmitted to the BPSK modulator [27]. The conditional BER for various PPM is $P_{LPPM} = \frac{1}{2}erfc\left(\frac{1}{2\sqrt{2}}\sqrt{\gamma\frac{L}{2}\log_2 L}\right)$ [28–31], for L-ary pulse position modulation (LPPM), one data symbol consists of L time slots, only one time slot is valid and the rest are zero, that is, L is the average length of the symbol [28,31]. According to Equations (2) and (8) in reference [11], the definite relationship of the conditional BERs between LPPM and L-ary pulse position modulation–binary phase shift keying–subcarrier intensity modulation (LPPM–BPSK–SIM) can be derived. Utilizing $Q(x) = \frac{1}{2}erfc\left(\frac{x}{\sqrt{2}}\right)$, the conditional BER of the LPPM–BPSK–SIM is derived as

$$P_{LPPM-BPSK-SIM} = Q\left(\frac{1}{4}\sqrt{\gamma L\log_2 L}\right). \tag{6}$$

The unconditional BER of various PPM–BPSK–SIM is obtained as

$$P_{LPPM-BPSK-SIM} = \int_0^\infty Q\left(\frac{1}{4}\sqrt{\gamma L\log_2 L}\right)f_\gamma(\gamma)d\gamma. \tag{7}$$

3.3. Average Capacity

According to the definition of the average channel capacity [32,33], $\overline{C} = E\left[\log_2(1+\gamma)\right]$, $E[\bullet]$ denotes the expectation operator, the expression for the average channel capacity of the hybrid system is given by:

$$C_{ave} = \int_0^{\infty} \log_2(1+\gamma) f_\gamma(\gamma) d\gamma = \frac{1}{\ln 2} \int_0^{\infty} \frac{1}{1+\gamma} F_\gamma^c(\gamma) d\gamma, \tag{8}$$

where $F_\gamma^c(\gamma) = 1 - F_\gamma(\gamma)$, $\frac{1}{1+\gamma} = G_{0,0}^{1,1}\left(x \middle| \begin{matrix} 0 \\ 0 \end{matrix}\right)$, substituting Equation (4) into Equation (8) and using the expression of the extended generalized bivariate Meijer's G function (EGBMGF) in reference [34]. The average channel capacity is obtained as

$$C_{ave} = \frac{2^{(\alpha+\beta)} A \sqrt{C}}{16\pi \ln(2)} \overline{\gamma_1} G_{0,0:0,0:1,6}^{0,0:1,1:6,0}\left(\begin{matrix} - \\ - \end{matrix} \middle| \begin{matrix} 0 \\ 0 \end{matrix} \middle| \begin{matrix} \frac{\xi^2+1}{2} \\ \kappa_1 \end{matrix} \middle| \overline{\gamma_1}, \frac{(\alpha\beta)^2 C}{16\overline{\gamma_2}} \right). \tag{9}$$

4. Results and Discussion

The performance of the mixed RF–FSO system with hybrid PPM–BPSK–SIM is shown in figures below, including outage probability performance, average BER, and average channel capacity. Note that, the RF link is assumed to obey Rayleigh distribution and we use Gamma–Gamma fading to describe the turbulence of the FSO link from weak to strong, the receiver employs IM/DD. The parameters for turbulence intensity are shown in Table 1 [35,36], the wavelength is 1550 nm, and the distance between the transmitter and the receiver is set to be 5 km.

Table 1. Parameters for different turbulence intensities.

Turbulence	α	β
Weak	11.6	10.1
Moderate	4.0	1.9
Strong	4.2	1.4

The outage probability of the hybrid RF–FSO system against average SNR per hop is shown in Figure 2, effects of atmospheric turbulence and pointing errors are taken into account. γ_{th} is set to be 10 dB, the average SNRs of the two different links are equal to each other (i.e., $\overline{\gamma_1} = \overline{\gamma_2} = \overline{\gamma}$). As seen, the decrease of atmospheric turbulence leads to the improvement of the outage performance. Moreover, taking into consideration the pointing error, the greater the parameter, ξ, the better the outage performance. For instance, when the outage probability is $P_{out} = 10^{-3}$, the system has a 6 dB decrease on average SNR from strong turbulence to weak turbulence while $\xi = 1$, but the difference value is less than 0.5 dB while $\xi = 5$. Therefore, a conclusion is reached that pointing error aggravates the influence of atmospheric turbulence on outage probability.

Comparison diagram of average BER of BPSK in the FSO system and the hybrid RF–FSO system is shown in Figure 3. The simulation of BPSK refers to reference [37]. It can be observed that by adopting the RF link, the hybrid RF–FSO system has an improvement in BER performance compared to the FSO system. The BER of BPSK in the FSO system is lower than that in hybrid RF–FSO system, and higher average SNR of the RF channel leads to greater improvement. In order to study the performance of the hybrid RF–FSO system, we set $\overline{\gamma_1} = \overline{\gamma_2} = \overline{\gamma}$ in the following.

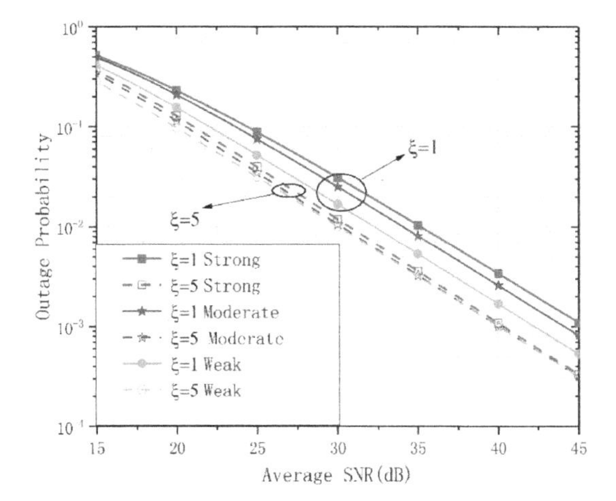

Figure 2. Outage probability against average signal-to-noise ratio (SNR) per hop for various turbulence intensities and different pointing error parameters. The constant is assumed to be $C = 1$, and the threshold of the hybrid system is $\gamma_{th} = 10$ dB.

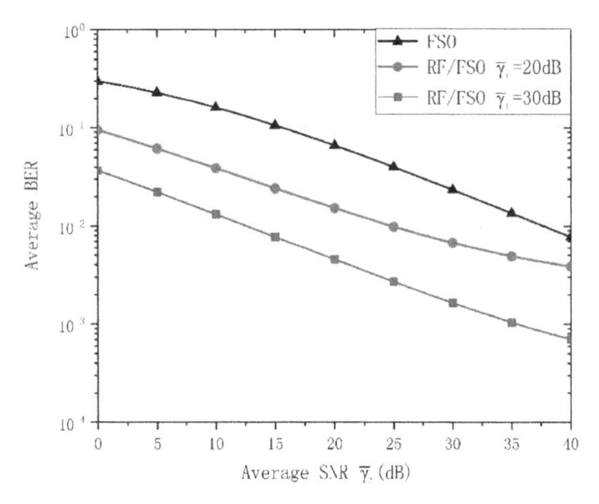

Figure 3. Average bit error rate (BER) of binary phase shift keying (BPSK) against average SNR under moderate turbulence and the pointing error parameter is $\xi = 1$. Turbulence parameters are $\alpha = 4.0$ and $\beta = 1.9$, the constant is assumed to be $C = 1$.

Figure 4 depicts the average BER of LPPM and BPSK in the mixed RF–FSO system against the average SNR per hop, the parameter of pointing error is 1 and $C = 1$. It can be seen from the figure that LPPM has a better BER performance as L increases. When the value of L is greater than four, the performance of LPPM in this system is better than that of BPSK. In Figure 5, comparison between various PPM–BPSK–SIM and BPSK has been shown, the same conclusion can be drawn that the BER of the hybrid system decreases when L increases, but the marginal values are different. For instance, when the value of L is greater than eight instead of four, the hybrid system adopting various PPM–BPSK–SIM achieves better performance than the system with BPSK. Simulation results in Figure 6 show that under moderate turbulence, a decrease in the value of C causes the improvement of the average BER, but it is irrelevant to the marginal value, the PPM–BPSK–SIM outperforms BPSK when the symbol length is greater than eight, that is, this is not a special case. An increase in the symbol length will reduce the bandwidth efficiency, which leads to an optimization of system performance for a reduction of average BER. Furthermore, the performance of the mixed modulation is affected by both modulations. Therefore, the value of L has a significant effect on the BER performance of hybrid system, the BER decreases as L increases, and the hybrid RF–FSO system achieves better BER performance by using the hybrid PPM–BPSK–SIM scheme only if the length of symbol, L, is greater than eight.

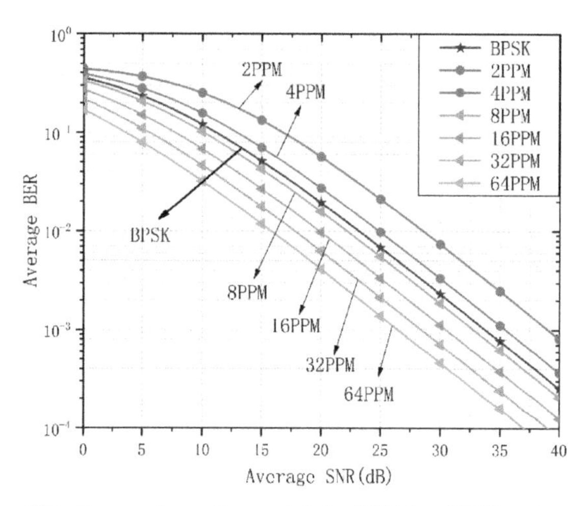

Figure 4. Average BER of L-ary pulse position modulation (LPPM) and BPSK against average SNR per hop under strong turbulence and the pointing error parameter is $\xi = 1$. Turbulence parameters are $\alpha = 4.2$ and $\beta = 1.4$, the constant is assumed to be $C = 1$.

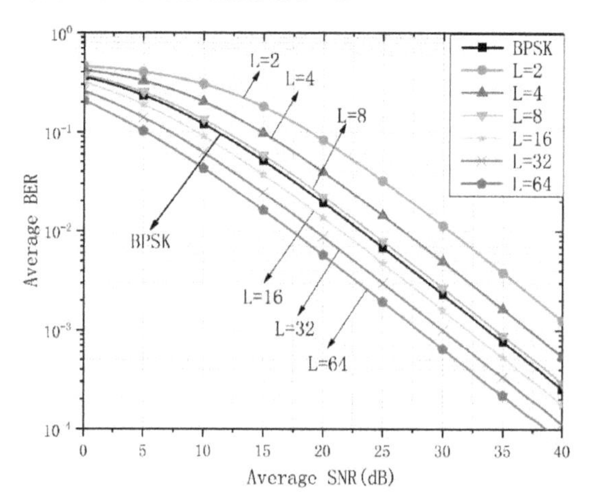

Figure 5. Average BER of L-ary pulse position modulation–binary phase shift keying–subcarrier intensity modulation (LPPM–BPSK–SIM) and BPSK against average SNR per hop under strong turbulence and the pointing error parameter is $\xi = 1$. Turbulence parameters are $\alpha = 4.2$ and $\beta = 1.4$, the constant is assumed to be $C = 1$.

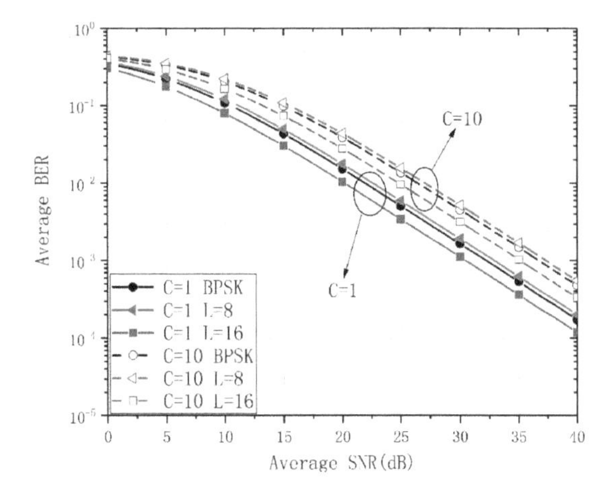

Figure 6. Average BER of LPPM–BPSK–SIM and BPSK against average SNR per hop under moderate turbulence and the pointing error parameter is $\xi = 1$. Turbulence parameters are $\alpha = 4.0$ and $\beta = 1.9$.

In Figure 7, the average BERs of LPPM–BPSK–SIM, LPPM, and BPSK in the hybrid RF–FSO system are presented under the same channel condition in Figure 4, we conducted studies about $L = 16$ and $L = 64$. It can be observed that when average SNR is fixed, the average BER of hybrid PPM–BPSK–SIM is higher than that of LPPM, and both of them outperform BPSK. For instance, when SNR = 25 dB, the average BER decreases in the order of BPSK, 16PPM–BPSK–SIM, 16PPM, 64PPM–BPSK–SIM, 64PPM. From another perspective, to achieve a BER of 10–3, 28 dB of SNR is required for 64PPM–BPSK–SIM and 32 dB is needed for 16PPM–BPSK–SIM, 33.5 dB is needed for BPSK. It can be noticed that the value of L has a great influence on the average BER of the system when using hybrid modulation, and the larger L, the lower the average BER. Considering the power efficiency and average BER of the hybrid modulation, we have conducted further research on the performance of 16PPM–BPSK–SIM.

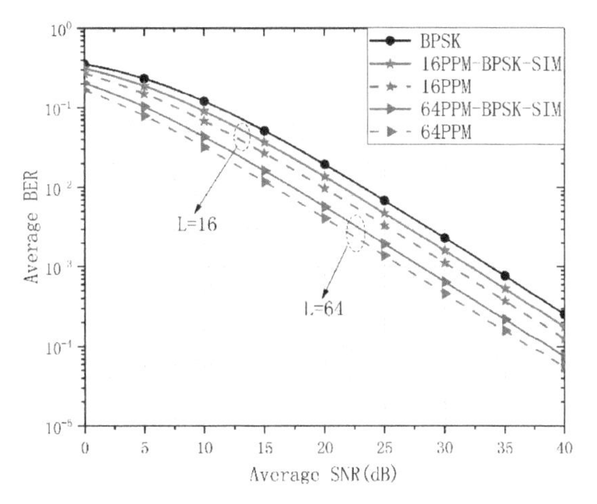

Figure 7. Average BER of hybrid modulation, LPPM, and BPSK against average SNR per hop under strong turbulence and the pointing error parameter is $\xi = 1$. Turbulence parameters are $\alpha = 4.2$ and $\beta = 1.4$, the constant is assumed to be $C = 1$.

In Figure 8, simulations of 16PPM–BPSK–SIM under different turbulence strength and different pointing error conditions have been implemented. What can be observed easily is that pointing error has a significant influence on the average BER of the hybrid system, the average BER increases as the value of ξ decreases. For example, regardless of the intensity of turbulence, the system with $\xi = 5$ outperforms the system with $\xi = 1$. On the other hand, when ξ is a fixed value, the average BER increases as the intensity of the turbulence increases, which is in line with our theoretical analysis. However, under moderate and weak turbulence with $\xi = 5$, the average BER of the system suddenly drops and appears to be undulating. After that, we changed the value of L and found that this condition still existed, this phenomenon indicates that the RF–FSO systematic performance is unstable. The adoption of hybrid PPM–BPSK–SIM has a stable improvement on average BER performance, especially under conditions with strong turbulence and severe pointing error.

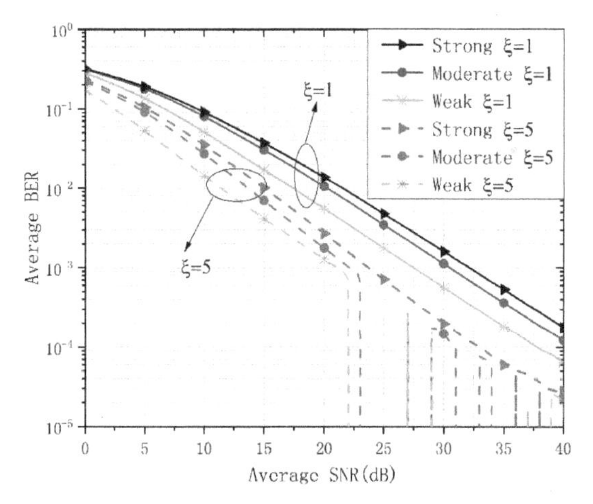

Figure 8. Average BER of 16PPM–BPSK–SIM against average SNR per hop for different turbulence strength and pointing error parameters. The fixed relay gain is $C = 1$.

In Figure 9, hybrid RF–FSO systematic average channel capacity against average SNR per hop is depicted with various turbulence intensities and different parameters of pointing error. As shown in Figure 9a, the average capacity of the system decreases with the increase of turbulence intensity regardless of the pointing error. When atmospheric turbulence strength is fixed, the larger the pointing error parameter is, the larger the system capacity will be. Further research is being carried out on the effect of pointing error on channel capacity. The differences of average capacity between strong and weak turbulence conditions are shown in Figure 9b. It can be seen from the graph that the smaller the pointing error parameter, the larger the maximum difference of average channel capacity. For instance, when $\xi = 10$, the maximum difference is 0.39 when SNR = 11 dB, but the maximum difference is 0.49 at SNR = 18 dB when $\xi = 1$. When the average SNR is greater than 11 dB, the channel capacity difference of $\xi = 1$ is larger than that at $\xi = 10$, though the gap between them will gradually shrink as the SNR value increases. We can conclude that a larger pointing error parameter results in a higher channel capacity, and it can aggravate the effect on capacity caused by atmospheric turbulence.

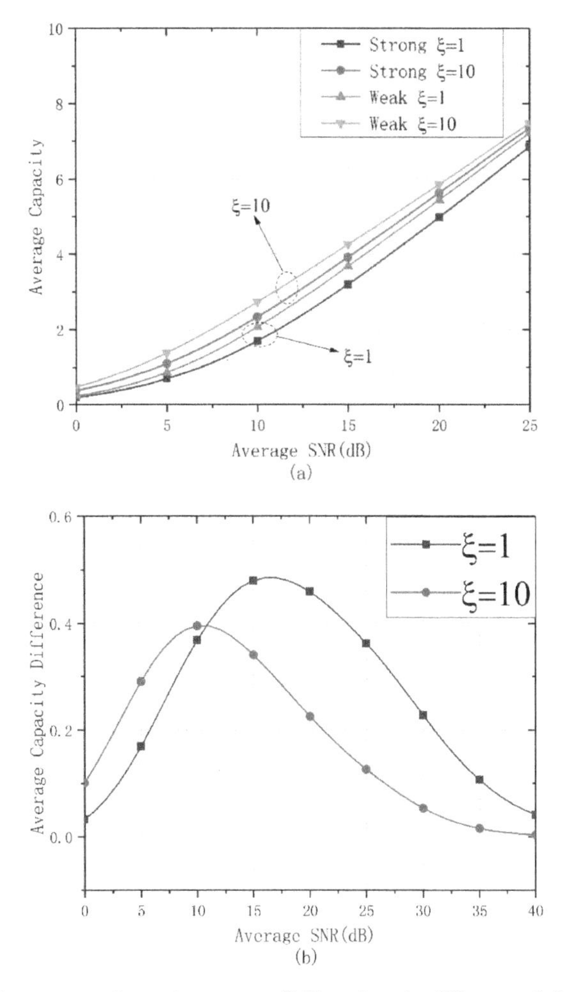

Figure 9. (**a**) Average capacity against average SNR per hop for different turbulence intensities and different pointing error parameters. (**b**) Average capacity difference between different pointing error parameters.

5. Conclusions

In summary, we studied the performance of mixed RF–FSO system with hybrid PPM–BPSK–SIM and fixed-gain relay. The RF link and the FSO link undergo Rayleigh distribution and Gamma–Gamma distribution severally. The expression of the unconditional BER of the hybrid RF–FSO system have been derived. On basis of the mathematical representation above, influences of the FSO link including atmospheric turbulence and pointing error have been investigated.

The results of our study show that the BER performance of the hybrid system could be ameliorated when it adopts hybrid PPM–BPSK–SIM, and the specific improvement effect has a lot to do with the length of symbol, L. The BER of the hybrid system degrades as the value of L increases, and when L is greater than eight, the hybrid system has a better BER performance than BPSK. The hybrid PPM–BPSK–SIM can improve the performance of the mixed RF–FSO system stably and obviously with strong turbulence and severe pointing error conditions. The outage probability and the average channel capacity are sensitive to turbulence effect and pointing error of the FSO channel. Simulation results also indicate that the pointing error can aggravate the effect of atmospheric turbulence on channel capacity and outage probability. Thus, the performance of mixed RF–FSO system could be improved by adopting hybrid PPM–BPSK–SIM, especially under conditions with strong turbulence influence and severe pointing error effect.

Author Contributions: Conceptualization: H.L.; formal analysis and writing: T.J.; investigation: L.Z.; methodology: X.Y.; validation: D.D. and A.L.; software: Z.W.

Funding: This study was funded by the National Natural Science Foundation of China; grant numbers are 61875057 and 61475049.

Conflicts of Interest: The authors declare no conflict of interest.

References

1. Lorences-Riesgo, A.; Guiomar, F.P.; Sousa, A.N.; Teixeira, A.L.; Muga, N.J.; Monteiro, P.P. 200 Gbit/s Free-Space Optics Transmission Using a Kramers-Kronig Receiver. In Proceedings of the Optical Fiber Communication Conference (OFC) 2019, San Diego, CA, USA, 3 March 2019.
2. Song, H.; Song, H.; Zhang, R.; Manukyan, K.; Li, L.; Zhao, Z.; Pang, K.; Liu, C.; Almaiman, A.; Bock, R.; et al. Experimental Mitigation of Atmospheric Turbulence Effect using Pre-Channel Combining Phase Patterns for Uni- and Bidirectional Free-Space Optical Links with Two 100-Gbit/s OAM-Multiplexed Channels. In Proceedings of the Optical Fiber Communication Conference Postdeadline Papers 2019, San Diego, CA, USA, 3 March 2019.
3. Gupta, J.; Dwivedi, V.K.; Karwal, V. On the Performance of RF-FSO System Over Rayleigh and Kappa-Mu/Inverse Gaussian Fading Environment. *IEEE Access* **2018**, *6*, 4186–4198. [CrossRef]
4. Anees, S.; Bhatnagar, M.R. Performance of an amplify-and-forward dual-hop asymmetric RF–FSO communication system. *J. Opt. Commun. Netw.* **2015**, *7*, 124–135. [CrossRef]
5. Samimi, H.; Uysal, M. End-to-end performance of mixed RF/FSO transmission systems. *IEEE/OSA J. Opt. Commun. Netw.* **2013**, *5*, 1139–1144. [CrossRef]
6. Zhang, J.; Dai, L.; Zhang, Y.; Wang, Z. Unified performance analysis of mixed radio frequency/free-space optical dual-hop transmission systems. *J. Lightwave Technol.* **2015**, *33*, 2286–2293. [CrossRef]
7. Zhang, Y.; Wang, X.; Zhao, S.-H.; Zhao, J.; Deng, B.-Y. On the performance of 2 × 2 DF relay mixed RF/FSO airborne system over Exponentiated Weibull fading channel. *Opt. Commun.* **2018**, *425*, 190–195. [CrossRef]
8. Petkovic, M.I.; Ansari, I.S.; Djordjevic, G.T.; Qaraqe, K.A. Error rate and ergodic capacity of RF-FSO system with partial relay selection in the presence of pointing errors. *Opt. Commun.* **2019**, *438*, 118–125. [CrossRef]
9. Zhao, J.; Zhao, S.-H.; Zhao, W.-H.; Liu, Y.; Li, X. Performance of mixed RF/FSO systems in exponentiated Weibull distributed channels. *Opt. Commun.* **2017**, *405*, 244–252. [CrossRef]
10. Torabi, M.; Effatpanahi, R. Performance analysis of hybrid RF-FSO systems with amplify-and-forward selection relaying. *Opt. Commun.* **2019**, *434*, 80–90. [CrossRef]
11. Faridzadeh, M.; Gholami, A.; Ghassemlooy, Z.; Rajbhandari, S. Hybrid PPM-BPSK subcarrier intensity modulation for free space optical communications. In Proceedings of the 2011 16th European Conference on Networks and Optical Communications (NOC), Newcastle, UK, 20–22 July 2011; pp. 36–39.
12. Giri, R.K.; Patnaik, B. BER analysis and capacity evaluation of FSO system using hybrid subcarrier intensity modulation with receiver spatial diversity over log-normal and gamma–gamma channel model. *Opt. Quantum Electron.* **2018**, *50*, 231. [CrossRef]
13. Ansari, I.S.; Yilmaz, F.; Alouini, M.-S. Impact of pointing errors on the performance of mixed RF/FSO dual-hop transmission systems. *IEEE Wirel. Commun. Lett.* **2013**, *2*, 351–354. [CrossRef]
14. Zedini, E.; Ansari, I.S.; Alouini, M.-S. Performance analysis of mixed Nakagami-*m* and Gamma–Gamma dual-hop FSO transmission systems. *IEEE Photonics J.* **2014**, *7*, 1–20. [CrossRef]
15. Prabu, K.; Kumar, D.S.; Srinivas, T. Performance analysis of FSO links under strong atmospheric turbulence conditions using various modulation schemes. *Opt. Int. J. Light Electron Opt.* **2014**, *125*, 5573–5581. [CrossRef]

16. Xu, F.; Lau, F.C.; Yue, D.-W. Diversity order for amplify-and-forward dual-hop systems with fixed-gain relay under Nakagami fading channels. *IEEE Trans. Wirel. Commun.* **2010**, *9*, 92–98. [CrossRef]
17. Aggarwal, M.; Garg, P.; Puri, P. Dual-hop optical wireless relaying over turbulence channels with pointing error impairments. *J. Lightwave Technol.* **2014**, *32*, 1821–1828. [CrossRef]
18. Ansari, I.S.; Yilmaz, F.; Alouini, M.-S. Performance analysis of FSO links over unified Gamma-Gamma turbulence channels. In Proceedings of the 2015 IEEE 81st Vehicular Technology Conference (VTC Spring), Glasgow, Scotland, 11–14 May 2015; pp. 1–5.
19. Bhatnagar, M.R.; Ghassemlooy, Z. Performance evaluation of FSO MIMO links in Gamma-Gamma fading with pointing errors. In Proceedings of the 2015 IEEE International Conference on Communications (ICC), London, UK, 8–12 June 2015; pp. 5084–5090.
20. Liu, C.; Yao, Y.; Sun, Y.; Zhao, X. Analysis of average capacity for free-space optical links with pointing errors over gamma-gamma turbulence channels. *Chin. Opt. Lett.* **2010**, *8*, 537–540.
21. Trung, H.D.; Pham, A.T. Pointing error effects on performance of free-space optical communication systems using SC-QAM signals over atmospheric turbulence channels. *AEU-Int. J. Electron. Commun.* **2014**, *68*, 869–876. [CrossRef]
22. Yang, L.; Gao, X.; Alouini, M.-S. Performance analysis of relay-assisted all-optical FSO networks over strong atmospheric turbulence channels with pointing errors. *J. Lightwave Technol.* **2014**, *32*, 4613–4620. [CrossRef]
23. Gradshteyn, I.S.; Ryzhik, I.M. *Table of Integrals, Series, and Products*; Academic Press: Cambridge, MA, USA, 2014.
24. Al-Eryani, Y.F.; Salhab, A.M.; Zummo, S.A.; Alouini, M.-S. Two-way multiuser mixed RF/FSO relaying: Performance analysis and power allocation. *IEEE/OSA J. Opt. Commun. Netw.* **2018**, *10*, 396–408. [CrossRef]
25. Adamchik, V.; Marichev, O. The algorithm for calculating integrals of hypergeometric type functions and its realization in REDUCE system. In Proceedings of the International Symposium on Symbolic and Algebraic Computation, Tokyo, Japan, 20–24 August 1990; pp. 212–224.
26. Hasna, M.O.; Alouini, M.-S. A performance study of dual-hop transmissions with fixed gain relays. *IEEE Trans. Wirel. Commun.* **2004**, *3*, 1963–1968. [CrossRef]
27. Faridzadeh, M.; Gholami, A.; Ghassemlooy, Z.; Rajbhandari, S. Hybrid pulse position modulation and binary phase shift keying subcarrier intensity modulation for free space optics in a weak and saturated turbulence channel. *JOSA A* **2012**, *29*, 1680–1685. [CrossRef]
28. Elganimi, T.Y. Performance comparison between OOK, PPM and pam modulation schemes for free space optical (FSO) communication systems: Analytical study. *Int. J. Comput. Appl.* **2013**, *79*. [CrossRef]
29. Faridzadeh, M.; Gholami, A.; Ghassemlooy, Z.; Rajbhandari, S. Hybrid 2-PPM-BPSK-SIM with the spatial diversity for free space optical communications. In Proceedings of the 2012 8th International Symposium on Communication Systems, Networks & Digital Signal Processing (CSNDSP), Poznan, Poland, 18–20 July 2012; pp. 1–5.
30. Trisno, S. Design and Analysis of Advanced Free Space Optical Communication Systems. Ph.D. Thesis, University of Maryland, College Park, MD, USA, 2006.
31. Yi, X.; Liu, Z.; Yue, P.; Shang, T. BER performance analysis for M-ary PPM over gamma-gamma atmospheric turbulence channels. In Proceedings of the 2010 6th International Conference on Wireless Communications Networking and Mobile Computing (WiCOM), Chengdu, China, 23–25 September 2010; pp. 1–4.
32. Si, C.; Zhang, Y.; Wang, Y.; Wang, J.; Jia, J. Average capacity for non-Kolmogorov turbulent slant optical links with beam wander corrected and pointing errors. *Opt. Int. J. Light Electron Opt.* **2012**, *123*, 1–5. [CrossRef]
33. Annamalai, A.; Palat, R.; Matyjas, J. Estimating ergodic capacity of cooperative analog relaying under different adaptive source transmission techniques. In Proceedings of the 2010 IEEE Sarnoff Symposium, Princeton, NJ, USA, 12–14 April 2010; pp. 1–5.
34. Ansari, I.S.; Al-Ahmadi, S.; Yilmaz, F.; Alouini, M.-S.; Yanikomeroglu, H. A new formula for the BER of binary modulations with dual-branch selection over generalized-K composite fading channels. *IEEE Trans. Commun.* **2011**, *59*, 2654–2658. [CrossRef]
35. Popoola, W.O.; Ghassemlooy, Z. BPSK subcarrier intensity modulated free-space optical communications in atmospheric turbulence. *J. Lightwave Technol.* **2009**, *27*, 967–973. [CrossRef]

36. Ghassemlooy, Z.; Popoola, W.; Rajbhandari, S. *Optical Wireless Communications: System and Channel Modelling with Matlab®*; CRC Press: Boca Raton, FL, USA, 2019.
37. Chatzidiamantis, N.D.; Karagiannidis, G.K.; Kriezis, E.E.; Matthaiou, M. Diversity combining in hybrid RF/FSO systems with PSK modulation. In Proceedings of the 2011 IEEE International Conference on Communications (ICC), Kyoto, Japan, 5–9 June 2011; pp. 1–6.

Article

Post-FEC Performance of Pilot-Aided Carrier Phase Estimation over Cycle Slip

Yan Li *, Quanyan Ning, Lei Yue, Honghang Zhou, Chao Gao, Yuyang Liu, Jifang Qiu, Wei Li, Xiaobin Hong and Jian Wu *

The State Key Laboratory of Information Photonics and Optical Communications, Beijing University of Posts and Telecommunications, Beijing 100876, China

* Correspondence: liyan1980@bupt.edu.cn (Y.L.); jianwu@bupt.edu.cn (J.W.)

Received: 27 June 2019; Accepted: 3 July 2019; Published: 8 July 2019

Abstract: The POST-forward error correction (FEC) bit error rate (BER) performance and the cycle-slip (CS) probability of the carrier phase estimation (CPE) scheme based on Viterbi–Viterbi phase estimation (VVPE) algorithm and the VV cascaded by pilot-aided-phase-unwrap (PAPU) algorithm have been experimentally investigated in a 56 Gbit/s quadrature phase-shift keying (QPSK) coherent communication system. Experimental results show that, with 0.78% pilot overhead, the VVPE + PAPU scheme greatly improves the POST-FEC performance degraded by continuous CS, maintaining a low CS probability with less influence of filter length. Comparing with the VVPE scheme, the VVPE + PAPU scheme can respectively obtain about 3.1 dB, 1.3 dB, 0.6 dB PRE-FEC optical signal noise ratio (OSNR) gains at PRE-BER of 1.8×10^{-2}. Meanwhile, the VVPE + PAPU scheme respectively achieves about 3 dB, 1 dB, and 0.5 dB POST-FEC OSNR gain and improves the FEC limit from 2.5×10^{-3} to 1.4×10^{-2}, from 8.9×10^{-3} to 1.8×10^{-2}, and from 1.6×10^{-2} to 1.9×10^{-2} under the CPE filter length of 8, 16, and 20.

Keywords: coherent communication; quadrature phase-shift keying; carrier phase estimation; cycle-slip; pilot-aided-phase-unwrap; low-density parity-check (LDPC)

---

1. Introduction

Recently, thanks to the advanced technology such as high-speed analog-to-digital converter (ADC), integrated optical front end, and digital signal processing (DSP), 100 Gb/s coherent single carrier transmission system has become a reality [1]. Single carrier quadrature phase-shift keying (QPSK) modulation and narrow line-width laser are a key technology in 100 Gbit/s digital coherent communication system. The laser line-width between 100 k and 500 k is commonly used. Carrier phase estimation (CPE) is an important integral part of digital signal processing (DSP) in coherent transmission systems through which laser phase noise and nonlinear phase noise are compensated [2,3]. Usually, CPE is implemented as a feed-forward structure for efficient hardware implementation, i.e., Viterbi–Viterbi phase estimation (VVPE) [4]. Due to the necessary phase unwrapping algorithm in these estimators, cycle slip (CS) occurs, which can lead to continuous errors that cannot be corrected properly by the forward error correction (FEC) decoder. A common approach to deal with CS is to use differential encoding (DE), however, differential decoding leads to error duplication and introduces penalties to the PRE- forward error correction (FEC) bit error rate (BER), thus leading to degraded POST-FEC BER [5].

A great deal of efforts has been conducted to mitigate or reduce the impact of CS and to improve the PRE-FEC BER and POST-FEC BER. A lot of research works focus on improved CPE algorithms, which can reduce the CS probability and decrease the PRE-FEC BER. For example, joint polarization carrier phase estimation (JP-CPE) [6,7], pilot-symbols-based CPE (P-CPE) [8], a forward and backward

(FWBW) method [9,10], recursive probability-weighted blind phase search (RW-BPS) [11], filtered carrier-phase estimation (F-CPE) [12], a support vector machine (SVM)-based boundary creation [13], and a signal recovery method by employing soft decision slip state estimation method at pilots [14]. However, most of these improved CPE algorithms mainly focus on the performance of PRE-FEC BER. The POST-FEC performance of the improved CPE algorithms in combination with a concatenated FEC coding scheme should be taken into consideration since CS might deteriorate the performance of FEC [5]. Some other works concentrate on improved FEC encoding/decoding methods, which can reduce differential encoding (DE)-penalty or improve robustness against CS. For example, turbo differential decoding (TDD) and some improved TDD algorithms have been proposed to eliminate error floor for frequent CS [15–18]. In References [19–21], novel code designs for BCH codes were proposed and numerically confirmed, which are robust to CS. Cao et al. [22] proposed a phase noise-aware log-likelihood ratio (LLR) calculation based on a Tikhonov model. An alternative LLR calculation method, employing linear or bilinear transform [23], was also investigated to improve robustness against residual phase noise. Schmalen [20] proposed a new low-density parity-check (LDPC) code structure, which is robust to CS. Clearly, these techniques, which treated CS within FEC coder/decoder block, can improve POST-FEC BER performance effectively. However, most of these techniques suffer from increased processing complexity.

In our previous work [24], the pilot-symbols-aided phase unwrapping (PAPU), which utilizes the time-division multiplexed pilot symbols that are transmitted with data, was proposed to do CS detection and correction with the CPE in QPSK modulation. Recently, it has been employed to high-order quadrature amplitude modulation (QAM) over 1500 km standard single-mode fiber transmission [25]. In this paper, since LDPC is one of the most commonly used FEC code [26], we experimentally investigate the performance of the CPE + PAPU scheme combined with a concatenated soft decision (SD)-LDPC coding in a 56 Gbit/s quadrature phase-shift keying (QPSK) coherent communication system with laser line-width of 300 kHz. Experimental results show that, compared with CPE, CPE + PAPU can release the required PRE-FEC optical signal noise ratio (OSNR) about 3.1 dB, 1.3 dB, and 0.6 dB at BER of 1.8×10^{-2} under the CPE filter length of 8, 16, and 20. Meanwhile, the CPE + PAPU scheme achieves respectively about 3 dB, 1 dB, and 0.5 dB POST-FEC OSNR gain compared with the CPE scheme under the CPE filter length of 8, 16, and 20. The results also show that the CPE + PAPU scheme respectively improve the FEC limit from 2.5×10^{-3}, 8.9×10^{-3}, 1.6×10^{-2} to 1.4×10^{-2}, 1.8×10^{-2}, 1.9×10^{-2} compared with CPE, with the CPE filter length of 8, 16, and 20.

2. Experimental Setup and Principles

A 28 Gbaud single-polarization (SP)-QPSK experiment has been conducted to investigate the POST-FEC performance and the CS probability of the previous proposed CPE + PAPU scheme [27]. Figure 1 illustrates the block diagram of the experimental setup. A DFB laser with 300-kHz line-width, centered at 1550 nm, is used as the transmitter laser. The independent pseudo-random binary sequences (PRBS) are encoded in sequence by outer encoder, Reed Solomon (RS) (255,239), and inner encoder LDPC. Inner encoder uses the most commonly used soft-decision code, DVB-S2 (LDPC), with 11.1% overhead [26] and its maximum decoder iteration number is 50. Pilot symbols with known information of 0.78% [27] overhead are inserted periodically per 127 symbols at the transmitter side and the phase estimated by the pilot symbols are used as a reliable reference in the CPE phase unwrap process at the receiver side. Pilot inserted data are loaded to pulse pattern generator (PPG) and then modulated by a double balanced modulator (IQ modulator) to generate 56 Gbit/s QPSK signals. An erbium-doped optical fiber amplifier (EDFA) cascaded by a variable optical attenuator (VOA) are used as noise source. An optical coupler (OC) with coupling ratio of 1:99 splits the output optical signals. One of branch is sent to the coherent receiver for signal processing, the other is used to detect optical signal noise ratio (OSNR). Agilent optical modulation analyzer with sampling rate of 80 GSa/s and bandwidth of 33 GHz is used as coherent receiver and off-line DSP. The key physical parameters in the experimental setup are specified and summarized in Table 1.

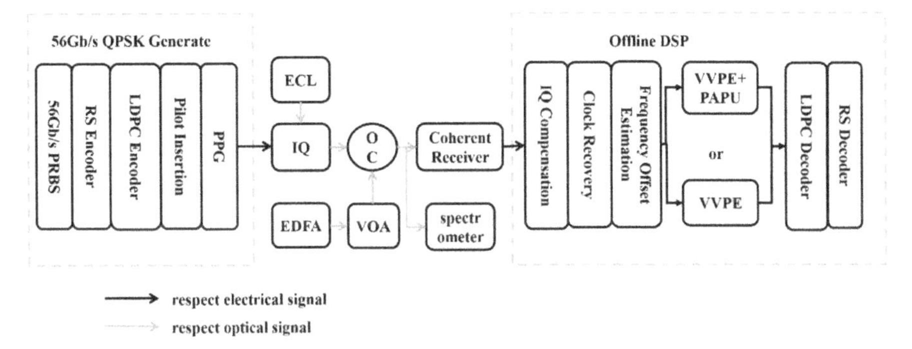

Figure 1. Experimental setup.

Table 1. Summary of the key physical parameters in experimental setup.

Parameters	Values
Symbol Rate	28-Gbaud
Wavelength	1550-nm
Linewidth	300-kHz
Bandwidth of Modulator	40-GHz
Bandwidth of Receiver	33-GHz
Sampling Rate	80-GSa/s

In the offline DSP, the signals are firstly processed by Gram–Schmidt orthogonalization procedure (GSOP) [28] to compensate in-phase (I) and quadrature (Q) imbalance. Then, the Gardner algorithm [29] is adopted to recovery clock and M-power frequency offset estimation (FOE) algorithm [30] is used to compensate frequency offset. Following the FOE, two different algorithms are employed to estimate the carrier phase. The first scenario is the traditional CPE based on VVPE, and the CPE + PAPU is adopted as the second scenario. The principle of the CPE scheme and the CPE + PAPU scheme are respectively shown in Figure 2a,b. The key point of the CPE + PAPU scheme, as shown in Figure 2b, is the pilot-symbols-aided phase unwrapping (PAPU), which does CS detection and correction in a completely forward way and combine itself with the usual phase unwrapping. Without loss of generality, we use VVPE as the CPE, and PAPU is implemented as the following procedure. After carrier phase recovery, LDPC decoder and RS decoder are conducted. Finally, the BER is calculated using 10 M bits data.

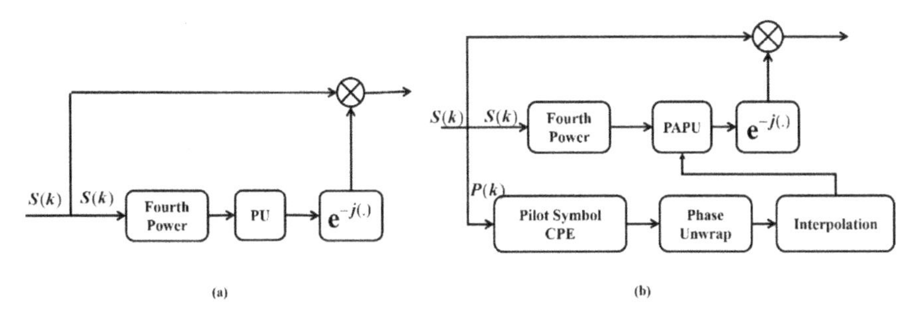

Figure 2. Fourth-power carrier phase recovery with (**a**) usual-phase unwrap, (**b**) pilot-symbol-aided phase unwrap.

3. Experimental Results and Discussion

Figure 3a–c show phase errors between the phase values estimated by CPE and the referenced phase values. Errors between CPE + PAPU and the referenced phase values are shown in Figure 3e–f. The traces in each figure indicate the errors under different CPE filter lengths (CPE filter length = 4, 8, 16, 20, 32, 48). To obtain the referenced phase values, we firstly calculate the phase error between original symbol and received symbol, then usual PU (2 pi) and an appropriate interpolation filter are used to remove the residual noise. The outputs of interpolation filter are used as referenced phase values. As shown in Figure 3, frequent CS is clearly observed for the VVPE scheme, producing $\pm 90^{\circ}$ phase error jumps under the condition of lower OSNR (which is equivalent to 10.5 dB or 12.1 dB) and small CPE filter length (which is less than 20). Even if OSNR is equivalent to 15.4 dB, the CPE scheme still shows frequent CS when CPE filter length is 4. Conversely, even when OSNR is equivalent to 10.5 dB and CPE filter length is less than 8, the VVPE + PAPU scheme just shows a few discrete CS, which are induced by additive white Gaussian noise (AWGN) and can be corrected by FEC. The results indicate that the CPE scheme seems to be more sensitive to CPE filter length and OSNR, compared to the CPE + PAPU scheme.

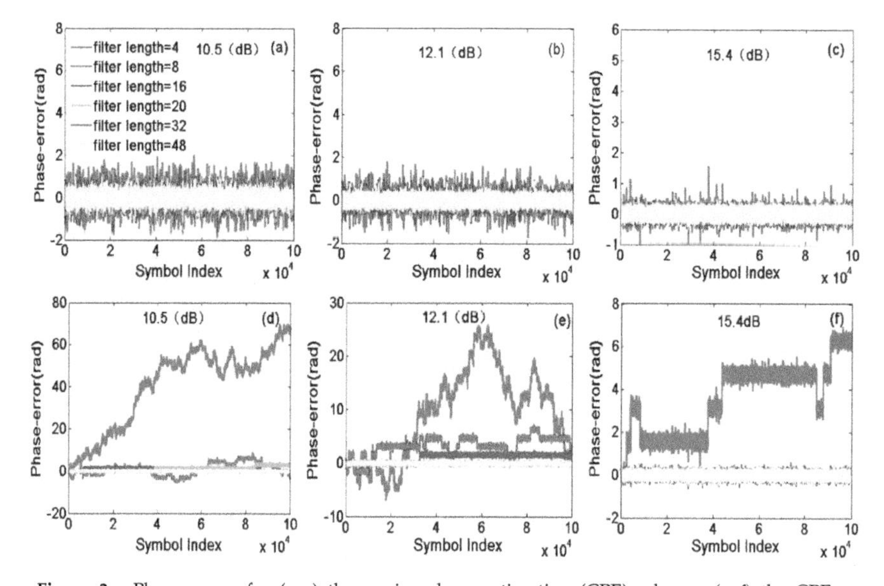

Figure 3. Phase errors for (**a–c**) the carrier phase estimation (CPE) scheme, (**e–f**) the CPE + pilot-aided-phase-unwrap (PAPU) scheme.

Probability of cycle slips as a function of OSNR under different CPE filter lengths are shown in Figure 4a. With the increasing of OSNR, the cycle slips probability of both schemes is decreasing and the difference between the VVPE scheme and the VVPE + PAPU scheme is becoming less obvious. The probability of CS is less than 1×10^{-7} when OSNR is large than 17.5 dB, which indicates that the CS problem can be avoided under high OSNR. However, with OSNR less than 12 dB, the VVPE scheme exhibits a high CS-rate with CPE filter length less than 48, compared with the VVPE + PAPU scheme. The reason is that frequent continuous CS occurs in the CPE scheme with lower OSNR, while only some discrete CS exists in the CPE + PAPU scheme. The results in Figure 4a indicate that the CPE + PAPU scheme can effectively mitigate the continuous CS due to the improper PU of the CPE. The discrete CS on account of AWGN cannot be corrected by the CPE + PAPU scheme, but these independent CS can be effectively corrected by FEC. Figure 4b shows the relationship between PRE-FEC BER and OSNR under different CPE filter lengths for the CPE scheme and the CPE + PAPU scheme. The AWGN

PRE-FEC BER is presented by gray circle. There is a similar trend between Figure 4a,b. When the CPE filter length is less than 48, the sudden degrade of PRE-FEC BER is consistent with the probability of CS for the CPE scheme.

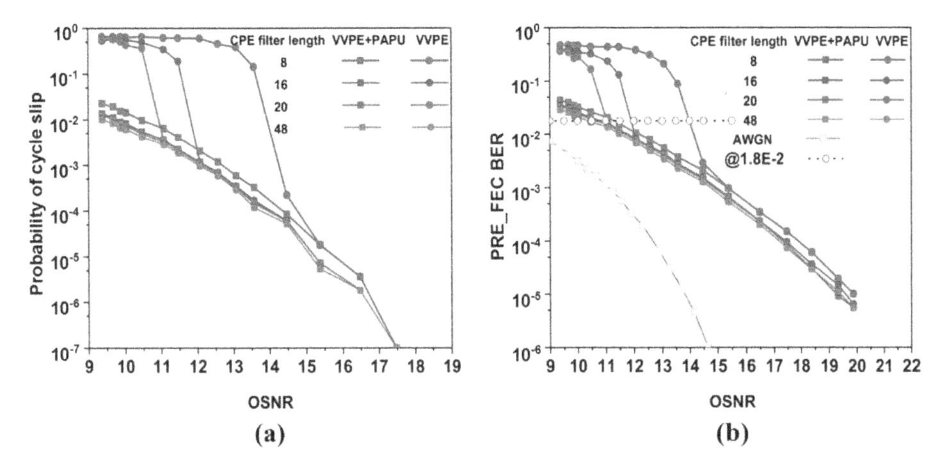

Figure 4. (**a**) Probability of cycle slips versus optical signal noise ratio (OSNR) for the CPE scheme and the CPE + PAPU scheme; (**b**) PRE-forward error correction (FEC) bit error rate (BER) versus OSNR for the CPE scheme and the CPE + PAPU scheme.

Figure 5a depicts the probability of CS versus CPE filter length under OSNR of 10.5 dB, 12.1 dB, and 15.4 dB for the CPE scheme and the CPE + PAPU scheme. As shown in Figure 5a, Under OSNR of 10.5 dB, 12.1 dB, and 15.4 dB, the CPE scheme exhibits high probability of CS with CPE filter length less than 32, 16, and 8, respectively. As shown in Figure 5b, with OSNR equivalent to 10.5 dB, 12.1 dB, 15.4 dB, to achieve the same PRE-FEC BER (respectively about 2×10^{-2}, 1×10^{-2}, 9.6×10^{-4}), the required CPE filter length of the CPE scheme and the CPE + PAPU scheme are 32 and 8, 16 and 6, and 8 and 8, respectively. With the OSNR equivalent to 10.5 dB, the PRE-FEC BER of CPE scheme sharply degrade from 2×10^{-2} to 0.5 with CPE filter length less than 32, moreover, the tendency of PRE-FEC BER agree with the CS probability in Figure 5a. When OSNR is increasing from 12.1 dB to 15.4 dB, the min CPE filter length to avoid the sharply degrading of PRE-FEC BER performance, decreases from 16 to 8.

Figure 6a shows the comparison of PRE-FEC BER and POST-FEC BER versus CPE filter lengths with different OSNR. When the OSNR is 10.5 dB, 12.1 dB, and 15.4 dB, the CPE filter lengths required to achieve 1×10^{-6} are 32, 16, 8 for CPE scheme and 20, 6, 2 for CPE + PAPU scheme. As shown in Figure 4, the probability of CS is decreasing and the performance of PRE-FEC BER is improving with the increasing of OSNR, thus the difference between the required CPE filter lengths to achieve 1×10^{-6} for two CPE schemes is small with high OSNR. With the lower OSNR, CPE + PAPU scheme is less sensitive to CPE filter length which may reduce the complexity of DSP.

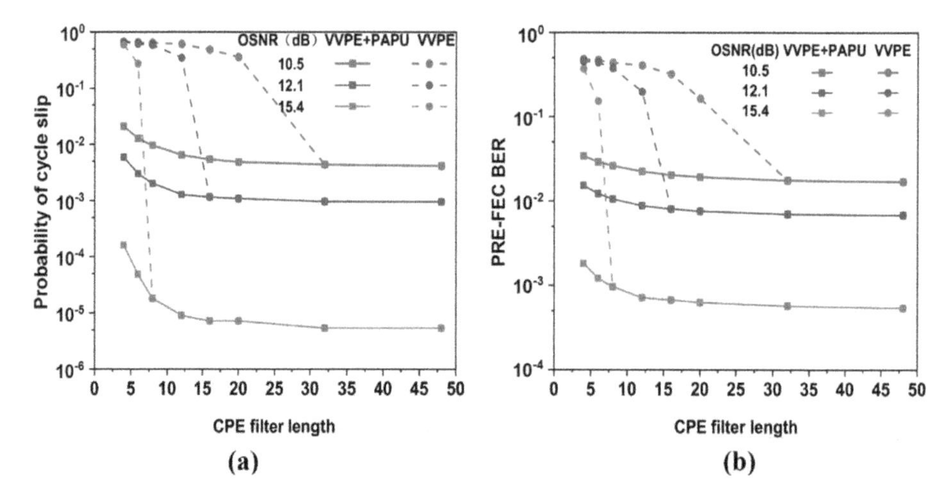

Figure 5. (**a**) Probability of cycle slips versus CPE filter length for the CPE scheme and the CPE + PAPU scheme; (**b**) PRE-FEC BER versus CPE filter length for the CPE scheme and the CPE + PAPU scheme, with OSNR equivalent to 10.5 dB, 12.1 dB, and 15.4 dB.

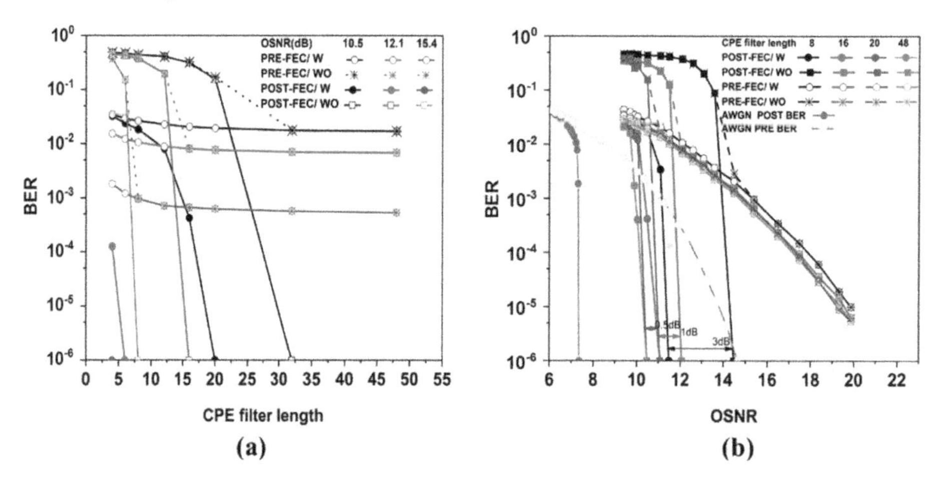

Figure 6. (**a**) BER versus CPE filter length curves of CPE and CPE + PAPU, with OSNR equivalent to 10.5 dB, 12.1 dB, and 15.4 dB; (**b**) BER versus OSNR curves of CPE and CPE + PAPU with different CPE filter lengths (W and WO respectively present the CPE +PAPU scheme and the CPE scheme).

The PRE-FEC BER and POST-FEC BER versus OSNR curves under different CPE filter lengths are shown in Figure 6b. Different CPE filter lengths are shown in the label. As is shown, the CPE + PAPU scheme outperforms the CPE scheme by about 3 dB at PRE-FEC BER of 2×10^{-2} with CPE filter length less than 16. It can be easily observed that slight degeneration of PRE-FEC BER may lead to rapid deterioration of POST-FEC BER due to CS. Figure 6b also depicts the POST-FEC BER performance of CPE and CPE + PAPU scheme indicated by square and circle connected with solid lines, respectively. The results prove that the CPE + PAPU scheme achieves about 3 dB, 1 dB, and 0.5 dB POST-FEC OSNR gain compared with the CPE scheme under the CPE filter length of 8, 16, and 20. The CPE + PAPU scheme outperforms the CPE scheme under shorter filter length benefitting from its effectively CS mitigation.

Figure 7 shows the relationship between POST-FEC BER and PRE-FEC BER on the CPE + PAPU scheme and the CPE scheme. The AWGN theory curve is represented by dark hollow circle. Apparently, CPE filter length has no significant effect on the BER transfer characteristic for CPE + PAPU scheme. When the CPE filter lengths are respectively equivalent to 8, 16, 20, and 48, the CPE + PAPU scheme successfully decodes signals with 1.4×10^{-2}, 1.8×10^{-2}, 1.9×10^{-2}, and 2.0×10^{-2} PRE-FEC BER and the CPE scheme fails to decode signals above 2.5×10^{-3}, 8.9×10^{-3}, 1.6×10^{-2}, and 2.0×10^{-2} PRE-FEC BER. As shown in Figure 4a, the CPE + PAPU scheme still present some independent CS, however, these CS can be effectively corrected by FEC. The continuous CS induced by inappropriate PU and amplifier spontaneous-emission (ASE) noise cannot be corrected by FEC. Comparing with the CPE scheme, the CPE + PAPU scheme seems to be less sensitive to CPE filter length. This might be attributed to the fact that the CPE + PAPU scheme effectively mitigates continuous CS induced by phase noise and inappropriate PU.

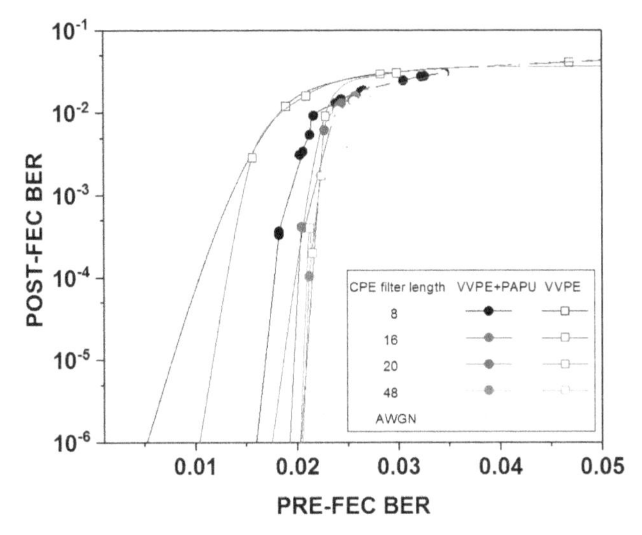

Figure 7. The relationship between POST-FEC BER and PRE-FEC BER.

4. Conclusions

In this paper, we have experimentally investigated the PRE-FEC BER and POST-FEC BER performance of the CPE + PAPU scheme and the CPE scheme on various CPE filter lengths in 56 Gbit/s QPSK systems. Experimental results show that, compared with CPE, CPE + PAPU can achieve the required PRE-FEC OSNR about 3.1 dB, 1.3 dB, and 0.6 dB at BER of 1.8×10^{-2} under the CPE filter length of 8, 16, and 20, meanwhile, the CPE + PAPU scheme respectively obtained 3 dB, 1 dB, and 0.5 dB POST-FEC OSNR gain and a large reduction of the CS rate compared with the CPE scheme, with the CPE filter length of 8, 16, and 20. The experimental results also prove that the CPE + PAPU scheme respectively improve the FEC limit from 2.5×10^{-3} to 1.4×10^{-2}, from 8.9×10^{-3} to 1.8×10^{-2}, and from 1.6×10^{-2} to 1.9×10^{-2} compared with the CPE scheme, with the CPE filter length of 8, 16, and 20. The experimental results show the CS-rate increases with deterioration of OSNR. On the lower OSNR region, CPE shows high residual CS at short filter length and low residual CS at long filter length, while CPE + PAPU can effectively reduce CS at all the filter lengths. Our experiment results prove that the CPE + PAPU scheme combined with a concatenated soft decision (SD)-LDPC coding can effectively reduce the CS induced by phase noise, which will be a promising candidate for next-generation high-speed coherent optical communications.

Author Contributions: This paper was mainly wrote by Y.L. (Yan Li) and Q.N., L.Y. demonstrated the experiment and H.Z. revised the article. C.G., Y.L. (Yuyang Liu), J.Q., W.L. and X.H. contributed to the reviewing and editing of the manuscript. J.W. supervised overall project.

Funding: This work was partly funded by 863 program 2015AA015503, NSFC program 61475022, 61505011, 61675034, 61331008, 973 project 2014CB340102, Fund of state key laboratory of IPOC (BUPT), and the fundamental research funds for the central universities.

Conflicts of Interest: The authors declare no conflict of interest.

References

1. Savory, S.J. Digital Coherent Optical Receivers: Algorithms and Subsystems. *IEEE J. Sel. Top. Quant. Electron.* **2010**, *16*, 1164–1178. [CrossRef]
2. Shieh, W.; Yang, Q.; Ma, Y. High-speed and high spectral efficiency coherent optical OFDM. In Proceedings of the Digest of the IEEE/LEOS Summer Topical Meetings, Acapulco, Mexico, 21–23 July 2008; pp. 115–116.
3. Lau, A.P.T.; Lu, C. Beyond 100 Gb/s: Advanced DSP techniques enabling high spectral efficiency and flexible optical communications. In Proceedings of the International Conference on Optical Communications and Networks (ICOCN), Chengdu, China, 26–28 July 2013; pp. 1–5.
4. Viterbi, A. Nonlinear estimation of PSK-modulated carrier phase with application to burst digital transmission. *IEEE Trans. Inf. Theory* **1983**, *4*, 543–551. [CrossRef]
5. Zhou, X. DSP for high spectral efficiency 400G transmission. In Proceedings of the European Conference and Exhibition on Optical Communication (2013), London, UK, 22–26 September 2013; pp. 1–3.
6. Gao, Y.; Lau, A.P.T.; Lu, C. Cycle-slip resilient carrier phase estimation for polarization multiplexed 16-QAM systems. In Proceedings of the Opto-Electronics and Communications Conference, Busan, Korea, 2–6 July 2012; pp. 154–155.
7. Bisplinghoff, A.; Vogel, C.; Kupfer, T.; Langenbach, S.; Schmauss, B. Slip-reduced carrier phase estimation for coherent transmission in the presence of non-linear phase noise. In Proceedings of the Optical Fiber Communication Conference and Exposition and the National Fiber Optic Engineers Conference (OFC/NFOEC), Anaheim, CA, USA, 17–21 March 2013; pp. 1–3.
8. Magarini, M.; Barletta, L.; Spalvieri, A.; Vacondio, F.; Pfau, T.; Pepe, M.; Bertolini, M.; Gavioli, G. Pilot-Symbols-Aided Carrier-Phase Recovery for 100-G PM-QPSK Digital Coherent Receivers. *IEEE Photonics Technol. Lett.* **2012**, *9*, 739–741. [CrossRef]
9. Zhang, H.; Cai, Y.; Foursa, D.G.; Pilipetskii, A.N. Cycle slip mitigation in POLMUX-QPSK modulation. In Proceedings of the Optical Fiber Communication Conference and Exposition and the National Fiber Optic Engineers Conference, Los Angeles, CA, USA, 6–10 March 2011; pp. 1–3.
10. Fludger, C.; Nuss, D.; Kupfer, T. Cycle-Slips in 100G DP-QPSK Transmission Systems. In Proceedings of the Optical Fiber Communication Conference (OFC), Los Angeles, CA, USA, 4–8 March 2012.
11. Rozental, V.N.; Kong, D.; Foo, B.; Corcoran, B.; Lowery, A. Low Complexity Blind Phase Recovery Algorithm with Increased Robustness Against Cycle-Slips. In Proceedings of the European Conference and Exhibition on Optical Communications (ECOC), Gothenburg, Sweden, 17–21 September 2017; pp. 1–3.
12. Rozental, V.N.; Kong, D.; Foo, B.; Corcoran, B.; Lowery, A. Cycle-slip-less low-complexity phase recovery algorithm for coherent optical receivers. *Opt. Lett.* **2017**, *18*, 3554–3557. [CrossRef] [PubMed]
13. Kawase, H.; Mori, Y.; Hasegawa, H.; Sato, K. Cycle-slip-tolerant decision-boundary creation with machine learning. In Proceedings of the International Conference on Photonics (ICP), Kuching, Malaysia, 14–16 March 2016; pp. 1–3.
14. Koike-Akino, T.; Yoshida, T.; Parsons, K.; Millar, D.S.; Kojima, K.; Pajovic, M. Fully-parallel soft-decision cycle slip recovery. In Proceedings of the Optical Fiber Communications Conference and Exhibition (OFC), Los Angeles, CA, USA, 19–23 March 2017.
15. Yu, F.; Stojanovic, N.; Hauske, F.N.; Chang, D.; Xiao, Z.; Bauch, G.; Pflueger, D.; Xie, C.; Zhao, Y.; Jin, L.; et al. Soft-decision LDPC turbo decoding for DQPSK modulation in coherent optical receivers. In Proceedings of the European Conference and Exhibition on Optical Communication (ECOC), Geneva, Switzerland, 18–22 September 2011.

16. Bisplinghoff, A.; Langenbach, S.; Kupfer, T.; Schmauss, B. Turbo Differential Decoding Failure for a Coherent Phase Slip Channel. In Proceedings of the European Conference and Exhibition on Optical Communication (ECOC), Amsterdam, The Netherlands, 16–20 September 2012.
17. Bisplinghoff, A.; Langenbach, S.; Beck, N.; Fludger, C.R.S.; Kupfer, T.; Schulien, C. Cycle slip tolerant hybrid turbo differential decoding. In Proceedings of the European Conference on Optical Communication (ECOC), Cannes, France, 21–25 September 2014; pp. 1–3.
18. Koike-Akino, T.; Kojima, K.; Millar, D.S.; Parsons, K.; Miyata, Y.; Matsumoto, W.; Sugihara, T.; Mizuochi, T. Cycle slip-mitigating turbo demodulation in LDPC-coded coherent optical communications. In Proceedings of the Optical Fiber Communications Conference and Exhibition (ECOC), San Francisco, CA, USA, 9–13 March 2014.
19. Leong, M.Y.; Larsen, K.J.; Jacobsen, G.; Popov, S.; Zibar, D.; Sergeyev, S. Novel BCH code design for mitigation of phase noise induced cycle slips in DQPSK systems. In Proceedings of the Lasers and Electro-Optics (CLEO), San Jose, CA, USA, 8–13 June 2014.
20. Schmalen, L. A Low-Complexity LDPC Coding Scheme for Channels with Phase Slips. *J. Light. Technol.* **2015**, *7*, 1319–1325. [CrossRef]
21. Schmalen, L. Low-complexity phase slip tolerant LDPC-based FEC scheme. In Proceedings of the European Conference on Optical Communication (ECOC), Cannes, France, 21–25 September 2014; pp. 1–3.
22. Cao, S.; Kam, P.Y.; Yu, C.; Cheng, X. Pilot-Tone Assisted Log-Likelihood Ratio for LDPC Coded CO-OFDM System. *IEEE Photonics Technol. Lett.* **2014**, *15*, 1577–1580. [CrossRef]
23. Koikeakino, T.; Millar, D.S.; Kojima, K.; Parsons, K. Phase Noise-Robust LLR Calculation with Linear/Bilinear Transform for LDPC-Coded Coherent Communications. In Proceedings of the Lasers and Electro-Optics (CLEO), San Jose, CA, USA, 10–15 May 2015.
24. Cheng, H.; Li, Y.; Zhang, F.; Wu, J.; Lu, J.; Zhang, G.; Xu, J.; Lin, J. Experimental Demonstration of Pilot-Symbols-Aided Cycle Slip Mitigation for QPSK Modulation Format. *Opt. Express* **2013**, *19*, 22166–22172. [CrossRef] [PubMed]
25. Lu, J.; Fu, S.; Hu, Z.; Deng, L.; Tang, M.; Liu, D.; Chan, C.C.K. Carrier Phase Recovery for Set-Partitioning QAM Formats. *J. Light. Technol.* **2018**, *18*, 4129–4137. [CrossRef]
26. DVB-S.2 Standard Specification, ETSI EN 302 307 V1.3.1. Available online: http://www.etsi.org/deliver/etsi_en/302300_302399/302307/01.03.01_60/en_302307v010301p.pdf (accessed on 1 March 2013).
27. Cheng, H.; Li, Y.; Kong, D.; Zang, J.; Wu, J.; Lin, J. Low overhead slipless carrier phase estimation scheme. *Opt. Express* **2014**, *17*, 20740–20747. [CrossRef] [PubMed]
28. Chang, S.H.; Chung, H.S.; Kim, K. Impact of quadrature imbalance in optical coherent QPSK receiver. *IEEE Photonics Technol. Lett.* **2009**, *11*, 709–711. [CrossRef]
29. Gardner, F. A BPSK/QPSK timing-error detector for sampled receivers. *IEEE Trans. Commun.* **1986**, *5*, 423–429. [CrossRef]
30. Leven, A.; Kaneda, N.; Koc, U.V.; Chen, Y.K. Frequency estimation in intradyne reception. *IEEE Photonics Technol. Lett.* **2007**, *6*, 366–368. [CrossRef]

Article

IBP Based Caching Strategy in D2D

Chun Shan, Xiao-ping Wu, Yan Liu *, Jun Cai and Jian-zhen Luo

School of Electronic and Information, Guangdong Polytechnic Normal University, Guangzhou 510665, China; shanchun@gpnu.edu.cn (C.S.); wxp3126129@163.com (X.-p.W.); gzhcaijun@126.com (J.C.); luojz@gpnu.edu.cn (J.-z.L.)

* Correspondence: liuyan_sysu@163.com

Received: 31 March 2019; Accepted: 27 May 2019; Published: 13 June 2019

Abstract: Device to Device (D2D) communication is a key technology in 5th generation wireless systems to increase communication capacity and spectral efficiency. Applying caching into D2D communication networks, the device can retrieve content from other devices by establishing D2D communication links. In this way, the backhaul traffic can be significantly reduced. However, most of the existing caching schemes in D2D are proactive caching, which cannot satisfy the requirement of real-time updating. In this paper, we propose an Indian Buffet Process based D2D caching strategy (IBPSC). Firstly, we construct a geographical D2D communication network to provide high quality D2D communications according to physical closeness between devices. Then devices are divided into several social communities. Devices are ranked by their node importance to community in each community. The base station makes caching decisions for devices according to contrition degree. Experimental results show that IBPSC achieves best network performance.

Keywords: Device to Device; caching; Indian Buffet Process

1. Introduction

According to Cisco VNI (Visual Networking Index) report, mobile data traffic will grow 7-fold from 2017 to 2022, and video traffic will be 79% of global mobile data traffic by 2022. It will be a big challenge to current wireless network. To reduce traffic and energy cost of backhaul links, D2D communication has been proposed. Within D2D communication network, device can communicate with nearby device directly by establishing D2D communication link between them. Another potential solution to this challenge is caching, which has been proved can improve network performance of several network scenarios [1,2]. In recent years, caching also has been applied into CN (Core Network) and RAN (Ratio Access Network) of 5th generation wireless systems (5G) to reduce backhaul traffic and network latency [3–7].

Recently, several studies have proved that applying caching strategy into D2D communication network can offload backhaul traffic and improve user experience [8,9]. In D2D caching network, each device is equipped with cache space for caching popular content. Device can receive its desired content from nearby devices instead of BS (Base Station) through one-hop or multi-hop D2D communication. Caching scheme is an important part of D2D caching network and impacts network performance. Most existing D2D caching schemes are proactive caching, i.e., pre-caching popular content to devices before the content being requested during off-peak time. The performance of proactive caching scheme depends on accuracy of prediction.

According to a report released by data provider iResearch, China's short video mark was valued at 14 billion yuan ($2 billion) in 2018. With the explosive growth of short video mobile apps, proactive caching strategies face a big challenge, since short videos are posed by users in any time, it is impossible to make accurate prediction and update caches in time. Therefore, we propose an IBP based reactive caching strategy. Zhang et al. proved that the influence of user's content selection by other users can be modeled by Indian Buffet Process (IBP) [10]. IBP defines a probability distribution over equivalence classes of sparse binary matrices with a finite number of rows and an unbounded number of columns [11]. In our model, we firstly construct a reliable D2D communication network according to geographical closeness between devices as in [10]. Then devices in the reliable D2D communication network are grouped into several social communities according to interaction between devices and ranked by their node importance to community. The caching decision is made by BS depending on contribution degree, which is calculated by BS according to IBP. The contributions of this paper are as follows:

- To provide high quality D2D communication links, a reliable D2D communication network is constructed according to physical closeness. Then devices are divided into several social communities according to their interactions and ranked by node importance.
- We propose an IBP based social-aware caching strategy. Within a community, BS makes caching decision for devices according to the contribution of caching content to other devices. Contribution degree is defined to measure the contribution and calculated from IBP.
- Finally, we study the performance of proposed strategy and compare it with other strategies in terms of cache hit ratio and sum rate. The simulation results show that IBPSC strategy achieves better performance, especially in the condition of small cache size and large parameter α.

The rest of the paper is organized as follows: We give an overview of related works in Section 2. Section 3 introduces an IBP based socially-aware D2D (IBPSC) caching strategy. Section 4 presents the simulation results. Finally, Section 5 concludes the paper.

2. Related Works

By deploying caching into D2D communication, QoE (Quality of Experience) and QoS (Quality of Service) of users can be significantly improved. The authors have proved that the D2D network with caching performs much better than other solutions [9]. They also proposed a Maximum Distance Separable coding based caching strategy for D2D, each device cache coded block with equal probability, the content can be recovered by obtaining enough coded blocks from other devices [12]. Naderializadeh et al. applied caching into a spectrum sharing mechanism D2D network to improve D2D spectral efficiency [13]. Golrezaei et al. divided the D2D caching network into several clusters, then introduced two caching strategies, deterministic caching strategy and Zipf-Based Random caching strategy. For deterministic caching strategy, each device caches the content with high popularity without duplication. For random caching, the distribution of contents cached by devices randomly and independently follows Zifp distribution [14]. Afshang et al. used Poisson cluster process for modeling locations of devices and proposed a cluster-centric content placement strategy to optimize the performance of the whole cluster [15]. Lee et al. proposed a clustering approach to optimize energy efficiency and throughout for based station assisted D2D caching network [16]. Giatsoglou et al. proposed a D2D-aware caching policy for millimeter-wave network to achieve higher offloading and lower content-retrieval delays [17]. Gregori et al. obtained optimal transmission and caching by formulating caching to a continuous time optimization problem [18]. The authors in [19] proposed a mobility aware caching strategy, the devices with low-speed and high-speed cache most popular content, the other devices cache content with lower popularity. Krishnan et al. proposed a Poisson Point Process based caching strategy for D2D to reduce latency, the locations of devices are modeled by Poisson Point Process, each device randomly caches a portion of content [20]. The authors proposed a probabilistic caching strategy for D2D caching network to maximize the cache hit rate and cache-aided throughput [21]. Chen et al. proposed a user-centric protocol and optimized proactive caching scheme to obtain high offloading gain with low energy cost at helper users [22]. Malak et al. designed a spatially correlated caching scheme, hard-core placement to increase cache hit rate. With the proposed scheme, the devices caching the same content are never closer to each other than the exclusion radius [23]. Recently, social relationship between users has been considered in D2D network. Zhu et al. proposed a socially-aware incentive approach to incentivize users to cache content for others and minimize total cost of the network for retrieving content [24]. A hypergraph framework is proposed by considering social ties, common interests and D2D transmission scheme for caching based D2D communications [25]. Wu et al. defined a notion of socially-aware rate, then proposed a socially-aware rate based content sharing mode selection strategy and modelled it as a maximum weighted mixed matching problem [26]. Wu et al. proposed a D2D relay-aided content caching strategy, which considers the different concentration levels of contents request from different users' groups and user's physical link [27].

In summary, most D2D caching schemes are proactive caching and cannot update cached content in time. Moreover, they require accurate prediction of content popularity in future. Several studies focusing on frameworks and incentive schemes are proposed, which have considered social relationships between devices. In this paper, we propose an India Buffet Process based socially-aware caching strategy to improve caching efficiency. Each device determines whether to cache the received content independently according to contribution degree.

3. Method

In this section, we propose an IBP based socially-aware caching strategy for D2D communication network. Physical closeness is used for constructing a reliable D2D communication network. Then the devices are divided into several social communities and ranked by node importance according to their interactions. Within a social community, IBP is used for modeling the influence of other users on user's selection of content. Caching decision is made by BS according to the contribution of caching the content to other devices within same community.

To provide high quality D2D communication links, a reliable D2D communication network is constructed according to physical closeness between devices. Physical closeness w_{ij} defined by Zhang et al. [10] is the probability of establishing a stable D2D communication link between device i and device j in future, and it ranges from 0 to 1. BS is responsible for collecting devices encounter history and calculating physical closeness w_{ij}. In our scheme, we consider the D2D communication network as an undirected graph G(V,E), V is the set of devices, E is the set of edges. The weight of edge between device i and device j is physical closeness w_{ij}. Then the reliable D2D communication network, G'(V,E) is constructed according to physical closeness. Within the reliable network, the weight of any edge is larger than a threshold w_{th}. In other word, the devices can establish high quality D2D links between them with high probability in future.

In our model, we take social relationships between devices into consideration, i.e., the interactions between devices in history. We construct an interaction matrix A_{ij} to represent the interactions of devices, $a_{ij} = 1$, if there was an interaction between device i and j in history; otherwise, $a_{ij} = 0$. Given the interaction matrix, the communities can be determined by the method proposed by Chauhan et al. [28], the importance of node k to community can be calculated according to Equation (1) [29].

$$P_k = \left(\sum_{i=1}^{c} \frac{v_{ik}^2}{v_i^T v_i}\right)/c \tag{1}$$

where c is the number of communities, v_i is an eigenvector of interaction matrix A, and v_{ik} is the k-th element of v_i. P_k ranges from 0 to 1. Within each community, the device with highest node importance degree will be label as u_1 (i.e., the first user) in its community.

The authors in [10] have proved that IBP can model the influence of other users on user's selection of content. For an IBP with K dishes and N, the customers are labeled in ascending order, $c_1, c_2, ..., c_N$. The number of dishes selected by a customer follows Poisson distribution with parameter α. The selection on dishes of customer c_j is influenced and only influenced by its prior customers, $c_1, c_2, ..., c_{j-1}$, where $j = 2, 3, ..., N$. The probability of customer c_1 selecting each dish is α/K. The probability of customer c_j selecting dish k, $p(z_{j,k} = 1 \mid \mathbf{z}_{\text{-}j,k})$, is given by the following:

$$p(z_{j,k} = 1 \mid \mathbf{z}_{\text{-}j,k}) = \begin{cases} m_k^{j-1}/j, & \mathbf{z}_{\text{-}j,k} \neq 0 \\ \alpha/(j \cdot K_0^j), & \mathbf{z}_{\text{-}j,k} = 0 \end{cases} \tag{2}$$

where $\mathbf{Z}$ is a $N \times k$ matrix to describe which customers select which dishes, if the customer c_n has selected dish k, $z_{nk} = 1$; otherwise,$z_{nk} = 0$. $\mathbf{z}_{-j,k}$ is the set of precious customers that have selected dish k. For each customer c_m belonging to $\mathbf{z}_{-j,k}$, we have $z_{mk} = 1$, and $1 \leqslant m < j$. m_k^{j-1} is the number of precious customers who have selected dish k. K_0^j is the number of dishes which are not tasted by precious customers.

In our model, within each community, devices are labeled in an ascending order according to their node importance P_k, the device with highest node importance is labeled as first user, u_1, i.e., first customer in IBP. By modeling the influence of other users on content selection, we define contribution degree $\overline{I_{ik}}$ to indicate the contribution of caching content k to device u_i to other users within same community. The contribution degree is given by :

$$\overline{I_{ik}} = \frac{1}{N-1} \sum_{\substack{j=1; \\ j \notin \mathbf{z}_{-j,k}}}^{N} p(z_{j,k} = 1 \mid \mathbf{z}_{-j,k}) \tag{3}$$

where $p(z_{j,k} = 1 \mid \mathbf{z}_{-j,k})$ is the conditional probability of user u_j requesting content k given by Equation (2). High contribution degree indicates content k will be requested by other devices with high probability.

When BS receives a request sent by device u_i, firstly checks whether the content is cached by other devices. If the D2D caching network cannot respond to this request, BS will serve the request; otherwise, BS locates content holder with highest physical closeness. BS calculates contribution degree $\overline{I_{ik}}$ according to Equation (3) and makes caching decisions for device u_i by setting caching flag fq ($fq \in \{0,1\}$). If $\overline{I_{ik}} >= I_{th}$, the content can be cached by device u_i, i.e., $fq = 1$; otherwise, $fq = 0$, the content will not be cached by device u_i, as described in Algorithm 1. The caching threshold I_{th} is a design number and ranges from 0 to 1.

Algorithm 1 IBPSC strategy

Input: G(V,E), w_{th}, I_{th};
Output: fq (caching flag, fq, indicates whether device can cache the content);
Initialize: fq = 0;

1: BS collects encounter and interaction information of devices within the cellular network, calculates closeness w_{ij}, constructs reliable D2D communication network G'(V,E) ;
2: grouping the devices into several social communities and calculating node importance to community of each device by Equation (1);
3: **for** each community **do**
4: labeling devices according to their node importance to community, i.e., $u_1, u_2, \ldots\ldots, u_N$;
5: **end for**
6: **BS receives a request for content** f **sent by device** u_i
7: BS calculates contribution degree $\overline{I_{if}}$ according to Equation (3);
8: **if** $\overline{I_{if}} >= I_{th}$ **then**
9: let $fq = 1$;
10: **else**
11: let $fq = 0$;
12: **end if**
13: **if** content f can be retrieved from other devices in the D2D caching network **then**
14: BS locates content holder with highest physical closeness and sends caching flag fq to the content holder;
15: establishing D2D communication to transmit content f with caching flag fq;
16: **else**
17: BS servers the request directly with content f and caching flag fq;
18: **end if**
19: **When device** u_i **receives content** f
20: **if** $fq \neq 0$ **then**
21: device u_i caches content f;
22: **end if**

4. Performance Evaluation

In this section, we investigate the performance of proposed strategy and compare it with other caching strategies for D2D:

- Maximal closeness: BS locates content holder with highest closeness to user [10]. All the content received by device are cached.
- Zipf-based Caching: It is a proactive caching strategy, files are pre-cached into devices according to Zipf distribution [14].
- Most popular caching strategy: It is also a proactive caching strategy, the most popular files are pre-cached into devices according to the predication of content popularity.

In our simulations, devices are randomly distributed on a surface covered by a BS, the radius of BS is 500 m. The max D2D communication distance ranges from 5 m to 50 m. BS is responsible for collecting users' encounter and interaction history, also locating content holder(s). In our simulation, cellular spectrum is used for both cellular and D2D communications, i.e., underlay inband D2D [30]. Content selections of devices are modeled by IBP. We use cache hit ratio and sum rate [10] to study the performance of four caching schemes. We define cache hit ratio as the rate of requests satisfied by other devices to all the requests sent by devices.

In our strategy, BS determines whether the content can be cached by device according to contribution degree $\overline{I_{ik}}$ and caching threshold I_{th}. Firstly, we investigate the influence of caching threshold I_{th} on cache hit ratio under different Poisson parameter α and cache size, as shown in Figures 1 and 2. Parameter α reflects user's preference; a large α indicates that devices request new content (has not been requested by precious devices) with high probability. Cache hit ratio increases with I_{th} increasing in the first stage, then decreases to zero. This is because, once threshold I_{th} exceeds contrition degrees of all users, none of content can be cached in D2D network.

In Figures 3 and 4, we investigate the performance of all the strategies with different parameter α. Parameter α reflects users preference, a large α means devices request new content with high probability, which is not cached by other devices. In this case, the traffic to BS will increase. As shown in Figure 3, with the increasing of parameter α, caching hit ratio of all strategies decreases. The proposed IBPSC strategies achieve highest cache hit ratio compared to other schemes, and the advantages of IBPSC are more obvious with larger α. With α increasing, the total number of content requested by users increases, therefore the sum rate increases as shown in Figure 4. The proposed IBPSC strategy achieves best performance thanks to its contribution degree based caching strategy. The content cached by devices will be requested by other devices with high probability in future.

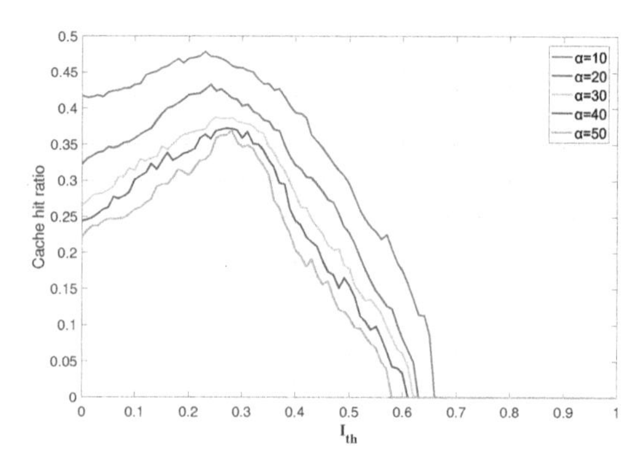

Figure 1. I_{th} with different Poisson parameter α.

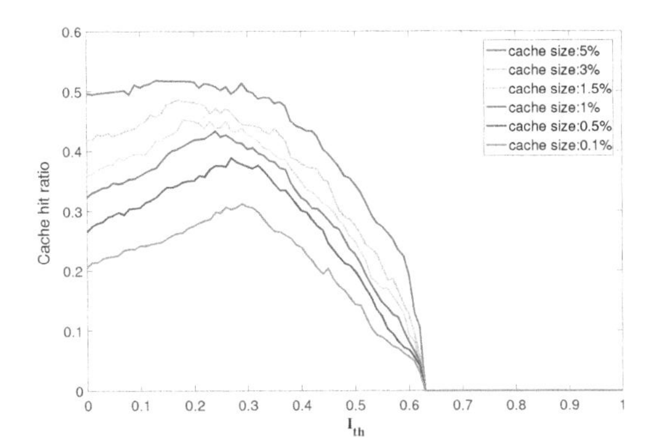

Figure 2. I_{th} with different cache size.

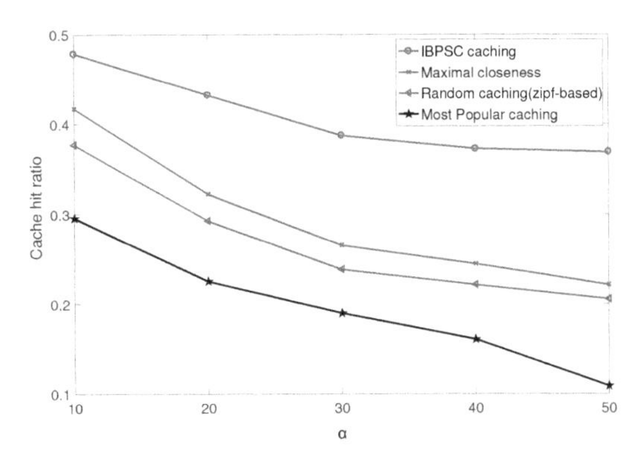

Figure 3. Cache hit ratio vs. Poisson parameter α.

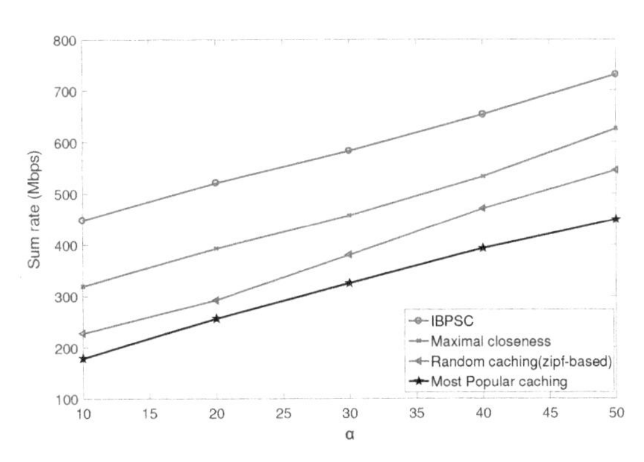

Figure 4. Sum rate vs. Poisson parameter α.

With the increasing of cache size, more content can be cached by devices, and more requests can be satisfied through D2D communications. Therefore, the cache hit ratio of all the caching strategies increases with the increasing of cache size as shown in Figure 5. Meanwhile, as more requests are satisfied by devices, the traffic to BS can be reduced. Figure 6 shows that the sum rate of all schemes increases as the cache size is increased. Since the proposed strategy takes probability of establishing

D2D communication in future into consideration, any two devices in the D2D communication network we constructed can establish a D2D communication in future with high probability. We also consider the social relationships between devices. Each device makes caching decision independently according to contribution to the community of caching the content. Only the content with high contribution degree can be cached by devices. Thus, the proposed IBPSC strategy performs much better than other three strategies in terms of caching hit ratio and sum rate, especially with small cache size.

Figures 7 and 8 show the performance of four strategies with different number of users. The cache hit ratio and sum rate increase with the number of users increasing, since more caching space is provided. With the increasing of Max D2D communication distance, the number of devices in D2D communication network increase, more caching space will be provided. Therefore the cache hit ratio and sum rate increase, as shown in Figures 9 and 10.

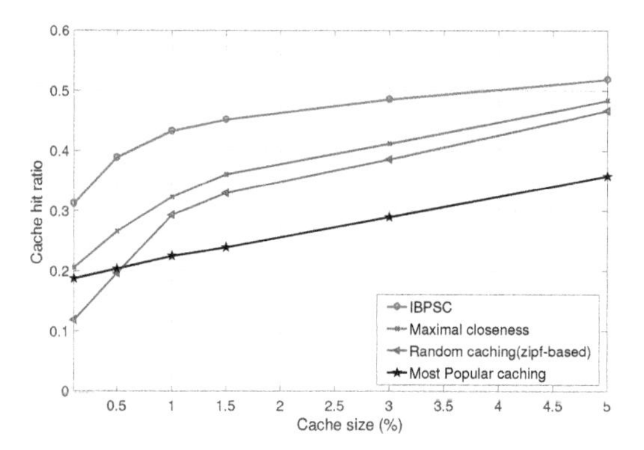

Figure 5. Cache hit ratio vs. cache size.

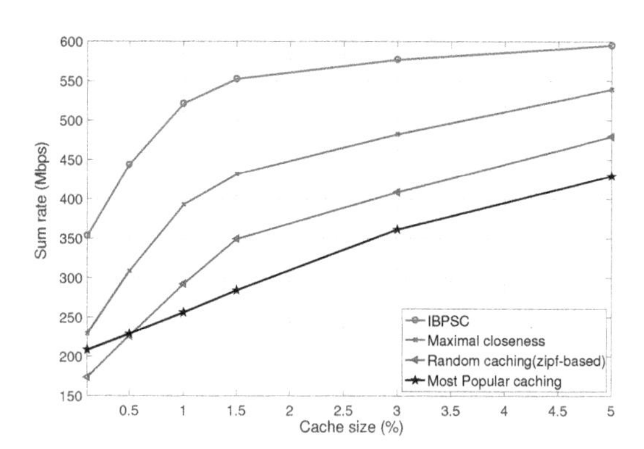

Figure 6. Sum rate vs. cache size.

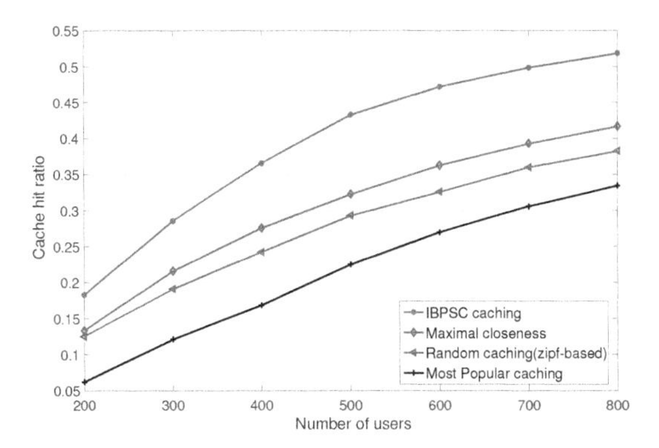

Figure 7. Cache hit ratio vs. number of users.

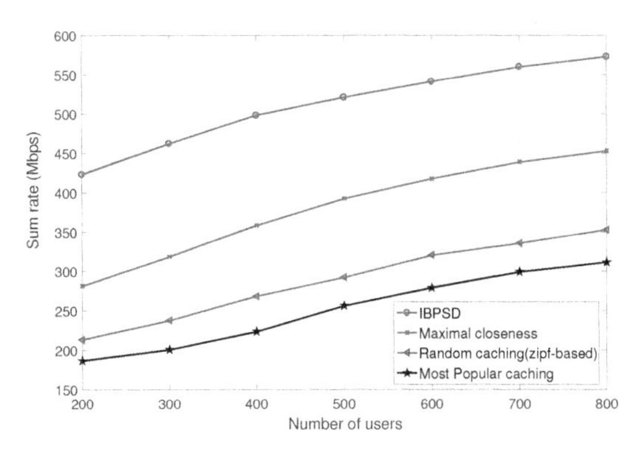

Figure 8. Sum rate vs. number of users.

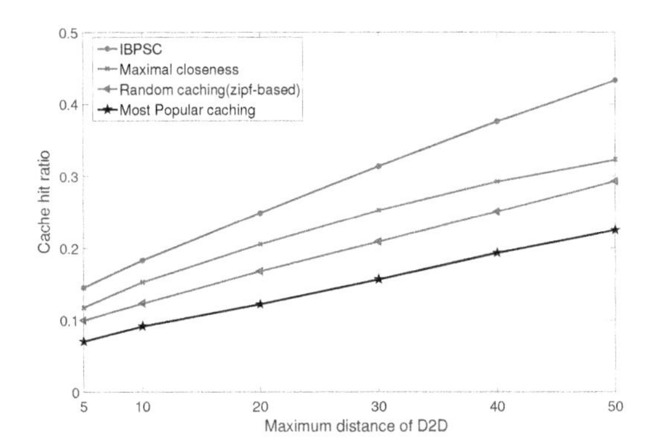

Figure 9. Cache hit ratio vs. maximum D2D communication distance.

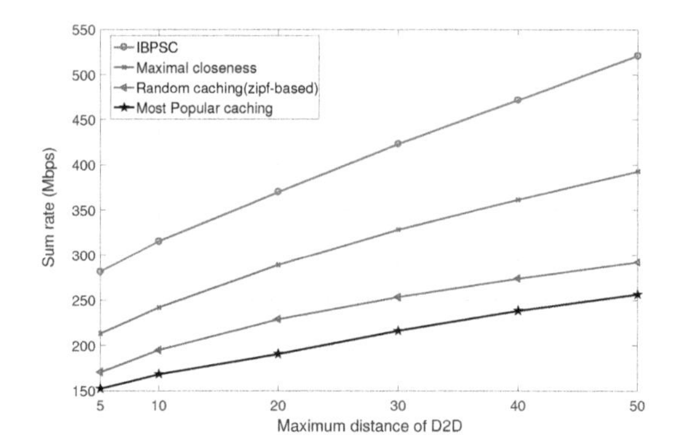

Figure 10. Sum rate vs. maximum D2D communication distance.

5. Conclusions

In this paper, we introduced an Indian Buffet Process based socially-aware D2D caching strategy (IBPSC). To provide high quality D2D communication links, a reliable D2D communication network is constructed according to physical closeness between devices. Then devices are grouped into several social communities and ranked by their node importance to community. BS calculates contribution degree and makes caching decisions for a device according to IBP. Finally, the experimental results have shown that the proposed IBPSC achieves highest network performance. In our future work, we intend to apply network coding into D2D caching network.

Author Contributions: Conceptualization and methodology, J.C., Y.L. and X.-p.W.; software, X.-p.W. and J.-z.L; writing—original draft preparation, Y.L. and X.-p.W.; writing—review and editing, Y.L. and J.C.; supervision, J.C. and C.S.

Funding: This research was funded by the National Natural Science Foundation of China (No.61571141, No.61702120); The Excellent Young Teachers in Universities in Guangdong (No.YQ2015105); Guangdong Provincial Application-oriented Technical Research and Development Special fund project (No.2015B010131017, No.2017B010125003); Guangdong Future Network Engineering Technology Research Center (No.2016GCZX006); Science and Technology Program of Guangzhou (No.201604016108); Science and Technology Project of Nan Shan (No.2017CX004); The Project of Youth Innovation Talent of Universities in Guangdong (No.2017KQNCX120); Guangdong science and technology development project (No.2017A090905023); The Key projects of Guangdong science and Technology (No.2017B030306015); The science and technology project in Guangzhou (No.201803010081).

Conflicts of Interest: The authors declare no conflict of interest.

Abbreviations

The following abbreviations are used in this manuscript:

CN	Core Network
RAN	Ratio Access Network
D2D	Device to Device
BS	Base Station
IBP	Indian Buffet Process

References

1. Borst, S.C.; Gupta, V.; Walid, A. Distributed Caching Algorithms for Content Distribution Networks. In Proceedings of the Conference on Information Communications, San Diego, CA, USA, 15–19 March 2010.
2. Din, I.U.; Hassan, S.; Khan, M.K.; Guizani, M.; Ghazali, O.; Habbal, A. Caching in Information-Centric Networking: Strategies, Challenges, and Future Research Directions. *IEEE Commun. Surv. Tutor.* **2018**, *20*, 1443–1474. [CrossRef]

3. Guo, K.; Yang, C.; Liu, T. Caching in Base Station with Recommendation via Q-Learning. In Proceedings of the Wireless Communications and NETWORKING Conference, San Francisco, CA, USA, 19–22 March 2017; pp. 1–6.
4. Vannithamby, R.; Talwar, S. Proactive Caching in 5G Small Cell Networks. In *Towards 5G: Applications, Requirements and Candidate Technologies*; Wiley: Hoboken, NJ, USA, 2017. [CrossRef]
5. Wang, X.; Chen, M.; Taleb, T.; Ksentini, A.; Leung, V.C.M. Cache in the air: Exploiting content caching and delivery techniques for 5G systems. *IEEE Commun. Mag.* **2014**, *52*, 131–139. [CrossRef]
6. Wang, X.; Li, X.; Leung, V.C.; Nasiopoulos, P. A framework of cooperative cell caching for the future mobile networks. In Proceedings of the 2015 48th Hawaii International Conference on System Sciences (HICSS), Kauai, HI, USA, 5–8 January 2015; pp. 5404–5413.
7. Parvez, I.; Rahmati, A.; Guvenc, I.; Sarwat, A.I.; Dai, H. A Survey on Low Latency Towards 5G: RAN, Core Network and Caching Solutions. *IEEE Commun. Surv. Tutor.* **2018**, *20*, 3098–3130. [CrossRef]
8. Golrezaei, N.; Molisch, A.F.; Dimakis, A.G.; Caire, G. Femtocaching and device-to-device collaboration: A new architecture for wireless video distribution. *IEEE Commun. Mag.* **2013**, *51*, 142–149. [CrossRef]
9. Ji, M.; Caire, G.; Molisch, A.F. Wireless Device-to-Device Caching Networks: Basic Principles and System Performance. *IEEE J. Sel. Areas Commun.* **2016**, *34*, 176–189. [CrossRef]
10. Zhang, Y.; Pan, E.; Song, L.; Saad, W.; Dawy, Z.; Han, Z. Social network aware device-to-device communication in wireless networks. *IEEE Trans. Wirel. Commun.* **2015**, *14*, 177–190. [CrossRef]
11. Griffiths, T.L.; Ghahramani, Z. The Indian Buffet Process: An Introduction and Review. *J. Mach. Learn. Res.* **2011**, *12*, 1185–1224.
12. Ji, M.; Caire, G.; Molisch, A.F. Fundamental Limits of Caching in Wireless D2D Networks. *IEEE Trans. Inf. Theory* **2016**, *62*, 849–869. [CrossRef]
13. Naderializadeh, N.; Kao, D.T.H.; Avestimehr, A.S. How to utilize caching to improve spectral efficiency in device-to-device wireless networks. In Proceedings of the 2014 52nd Annual Allerton Conference on Communication, Control, and Computing (Allerton), Monticello, IL, USA, 30 September–3 October 2014; pp. 415–422. [CrossRef]
14. Golrezaei, N.; Mansourifard, P.; Molisch, A.F.; Dimakis, A.G. Base-station assisted device-to-device communications for high-throughput wireless video networks. *IEEE Trans. Wirel. Commun.* **2014**, *13*, 3665–3676. [CrossRef]
15. Afshang, M.; Dhillon, H.S.; Chong, P.H.J. Fundamentals of cluster-centric content placement in cache-enabled device-to-device networks. *IEEE Trans. Commun.* **2016**, *64*, 2511–2526. [CrossRef]
16. Lee, M.C.; Molisch, A.F.; Ming, H. Caching Policy and Cooperation Distance Design for Base Station Assisted Wireless D2D Caching Networks: Throughput and Energy Efficiency Optimization and Trade-Off. *IEEE Trans. Wirel. Commun.* **2018**, *17*, 7500–7514. [CrossRef]
17. Giatsoglou, N.; Ntontin, K.; Kartsakli, E.; Antonopoulos, A.; Verikoukis, C. D2D-Aware Device Caching in mmWave-Cellular Networks. *IEEE J. Sel. Areas Commun.* **2017**, *35*, 2025–2037. [CrossRef]
18. Gregori, M.; Gómez-Vilardebó, J.; Matamoros, J.; Gündüz, D. Wireless Content Caching for Small Cell and D2D Networks. *IEEE J. Sel. Areas Commun.* **2016**, *34*, 1222–1234. [CrossRef]
19. Wang, R.; Zhang, J.; Song, S.H.; Letaief, K.B. Mobility-Aware Caching in D2D Networks. *IEEE Trans. Wirel. Commun.* **2017**, *16*, 5001–5015. [CrossRef]
20. Krishnan, S.; Dhillon, H.S. Distributed caching in device-to-device networks: A stochastic geometry perspective. In Proceedings of the 2015 49th Asilomar Conference on Signals, Systems and Computers, Pacific Grove, CA, USA, 8–11 November 2015; pp. 1280–1284. [CrossRef]
21. Chen, Z.; Pappas, N.; Kountouris, M. Probabilistic Caching in Wireless D2D Networks: Cache Hit Optimal Versus Throughput Optimal. *IEEE Commun. Lett.* **2017**, *21*, 584–587. [CrossRef]
22. Chen, B.; Yang, C.; Molisch, A.F. Cache-Enabled Device-to-Device Communications: Offloading Gain and Energy Cost. *IEEE Trans. Wirel. Commun.* **2017**, *16*, 4519–4536. [CrossRef]
23. Malak, D.; Al-Shalash, M.; Andrews, J.G. Spatially Correlated Content Caching for Device-to-Device Communications. *IEEE Trans. Wirel. Commun.* **2018**, *17*, 56–70. [CrossRef]
24. Zhu, K.; Zhi, W.; Zhang, L.; Chen, X.; Fu, X. Social-Aware Incentivized Caching for D2D Communications. *IEEE Access* **2016**, *4*, 7585–7593. [CrossRef]

25. Bai, B.; Wang, L.; Han, Z.; Chen, W.; Svensson, T. Caching based socially-aware D2D communications in wireless content delivery networks: A hypergraph framework. *IEEE Wirel. Commun.* **2016**, *23*, 74–81. [CrossRef]
26. Wu, D.; Zhou, L.; Cai, Y. Social-Aware Rate Based Content Sharing Mode Selection for D2D Content Sharing Scenarios. *IEEE Trans. Multimedia* **2017**, *19*, 2571–2582. [CrossRef]
27. Wu, K.; Jiang, M.; She, F.; Chen, X. Relay-aided Request-aware Distributed Packet Caching for Device-to-Device Communication. *IEEE Wirel. Commun. Lett.* **2018**, *8*, 217–220. [CrossRef]
28. Chauhan, S.; Girvan, M.; Ott, E. Spectral properties of networks with community structure. *Phys. Rev. E* **2009**, *80*, 056114. [CrossRef] [PubMed]
29. Wang, Y.; Di, Z.; Fan, Y. Identifying and characterizing nodes important to community structure using the spectrum of the graph. *PLoS ONE* **2011**, *6*, e27418. [CrossRef]
30. Asadi, A.; Wang, Q.; Mancuso, V. A survey on device-to-device communication in cellular networks. *IEEE Commun. Surv. Tutor.* **2014**, *16*, 1801–1819. [CrossRef]

MDPI
St. Alban-Anlage 66
4052 Basel
Switzerland
Tel. +41 61 683 77 34
Fax +41 61 302 89 18
www.mdpi.com

Applied Sciences Editorial Office
E-mail: applsci@mdpi.com
www.mdpi.com/journal/applsci

CPSIA information can be obtained
at www.ICGtesting.com
Printed in the USA
LVHW070016290520
656811LV00007B/454